T0205896

A Geometrical Approach to Physics

This book provides an accessible introduction to using the tools of differential geometry to tackle a wide range of topics in physics, with the concepts developed through numerous examples to help the reader become familiar with the techniques.

Physical applications are used to develop the techniques and demonstrate their wide-ranging applicability. Formalism is introduced sparingly and step-by-step, where it is needed, and chapters contain exercises for readers to test their understanding. Worked solutions to the exercises are included.

It is an ideal textbook for advanced undergraduate or postgraduate courses on mathematical methods for physicists, for students whose background is in physics rather than mathematics. It is assumed that the reader has no prior knowledge of mathematical methods beyond the content of a standard undergraduate physics degree.

The purpose of the book is to act as a 'gateway' to more advanced books on the applications of differential geometry in physics. It will also help the reader to better appreciate modern physics research that makes use of differential geometry, and the common features that permeate the discipline as a whole.

Key Features:
- Presents a light and accessible treatment.

- Can be used as a textbook for a short course on mathematical methods for physicists.

- Accessible to advanced undergraduates and postgraduates whose background is in physics, not mathematics.

David A. Burton received his PhD from Lancaster University, UK, in 2000 and was appointed Lecturer in Physics there in 2005. He is currently Senior Lecturer in Physics at Lancaster. He began his research career in relativistic continuum mechanics and gravitational physics before turning to fluid-structure interactions (in particular, vortex-induced vibration) and, in later years, to relativistic laser-plasma interactions.

Adam Noble received his PhD in 2006, also from Lancaster University, and has since held postdoctoral positions at Lancaster and the University of Strathclyde, Scotland, where he is currently a Research Fellow. His interests lie at the interface of physics with geometry, in particular electrodynamics of intense fields, plasma physics and particle physics.

The authors maintain a long-standing collaboration and, over the years, have worked on a number of topics connecting electromagnetics, gravitation and plasma physics, including gravitational Sagnac interferometry, relativistic wave-breaking in plasmas, radiation reaction in relativistic plasmas and charged particle beams, and the use of laser-wakefield accelerators in searches for light, weakly-interacting, candidates for dark matter.

A Geometrical
Approach to Physics

David A. Burton and Adam Noble

CRC Press
Taylor & Francis Group
Boca Raton London New York

CRC Press is an imprint of the
Taylor & Francis Group, an **informa** business

Designed cover image: CHIEW/Shutterstock

First edition published 2024
by CRC Press
2385 NW Executive Center Drive, Suite 320, Boca Raton FL 33431

and by CRC Press
4 Park Square, Milton Park, Abingdon, Oxon, OX14 4RN

CRC Press is an imprint of Taylor & Francis Group, LLC

ISBN: 978-1-032-13380-5 (hbk)
ISBN: 978-1-032-12928-0 (pbk)
ISBN: 978-1-003-22894-3 (ebk)

DOI: 10.1201/9781003228943

Typeset in LM Roman
by KnowledgeWorks Global Ltd.

Publisher's note: This book has been prepared from camera-ready copy provided by the authors.

To our teachers,
our students,
and our families.

Contents

Foreword   xiii

CHAPTER   1 ▪ Differential forms   1

1.1   MOTIVATION   1

1.2   ASPECTS OF THERMODYNAMICS IN TERMS OF DIFFERENTIAL FORMS   3

1.3   THE RULES OF EXTERIOR CALCULUS   6

CHAPTER   2 ▪ Vector fields and their relationship with differential forms   8

2.1   INTRODUCTION   8

2.2   DIFFERENTIAL 1-FORMS AS COTANGENT VECTOR FIELDS   9

    2.2.1   Tangent vector fields   9

    2.2.2   Heat capacity from a geometrical perspective   10

    2.2.3   Heat capacity and the first law of thermodynamics   12

2.3   INTERIOR PRODUCT ON DIFFERENTIAL FORMS   15

    2.3.1   A relationship between thermodynamic properties of a gas   15

2.4   CONSOLIDATION   16

    2.4.1   Differential forms   17

    2.4.2   The exterior derivative   18

| | 2.4.3 | Vector fields | 18 |
| | 2.4.4 | The interior product | 20 |

CHAPTER 3 ▪ Aspects of integration — 21

3.1	INTRODUCTION		21
3.2	A UNIFIED PERSPECTIVE ON MULTIVARIABLE INTEGRATION		21
	3.2.1	The emergence of 2-forms in the context of magnetostatics	21
	3.2.2	The relationship between integrals of differential forms and ordinary integrals	24
	3.2.3	The generalised Stokes theorem	27
3.3	AN ALTERNATIVE PERSPECTIVE ON TWO WELL-KNOWN IDENTITIES IN VECTOR CALCULUS		28

CHAPTER 4 ▪ The metric tensor — 30

4.1	INTRODUCING THE METRIC TENSOR		30
4.2	ORTHONORMAL FRAMES		33
4.3	MEASUREMENTS		37
	4.3.1	Lengths	37
	4.3.2	Angles and rapidities	39
	4.3.3	Volumes	41
4.4	EXAMPLES		43
	4.4.1	Hyperbolic space	43
	4.4.2	Cosmological spacetimes	44
	4.4.3	Rindler spacetime	45
4.5	DUALITIES		46
	4.5.1	Metric dual	46
	4.5.2	Hodge map	47
4.6	CONFORMAL STRUCTURE		50
4.7	SYMMETRIES AND KILLING VECTORS		52

CHAPTER   5 ▪ Maxwell's equations in terms of differential forms 56

5.1   INTRODUCTION   56
5.2   THE IMPORTANCE OF THE METRIC   56
5.3   THE VACUUM MAXWELL EQUATIONS FROM A
       FOUR-DIMENSIONAL PERSPECTIVE   58
5.4   ELECTROMAGNETIC WAVES FROM A SPACETIME
       PERSPECTIVE   62
5.5   OBSERVERS AND THE FIELDS THEY PERCEIVE   65
5.6   ELECTRIC CHARGE AND ELECTRIC CURRENT FROM
       A FOUR-DIMENSIONAL PERSPECTIVE   67
       5.6.1   Charge conservation   68
5.7   POLARISATION AND MAGNETISATION   72

CHAPTER   6 ▪ Classical mechanics   76

6.1   THE TANGENT BUNDLE   76
6.2   LAGRANGIAN MECHANICS   79
       6.2.1   Example: Free particle   81
       6.2.2   Example: Harmonic oscillator   82
       6.2.3   Example: Particle in a magnetic field   83
6.3   THE COTANGENT BUNDLE   84
6.4   HAMILTONIAN MECHANICS   87
6.5   POISSON BRACKETS   90
6.6   CONSERVED QUANTITIES   92
6.7   SINGULAR LAGRANGIANS   94
6.8   ALMOST SYMPLECTIC GEOMETRY   97

CHAPTER   7 ▪ Connections   100

7.1   INTRODUCTION   100
7.2   COVARIANT DIFFERENTIATION   101
7.3   PARALLEL TRANSPORT   103

7.4   GEODESICS AND THE LEVI-CIVITA CONNECTION   105

7.5   TORSION AND NONMETRICITY   107

     7.5.1   Torsion   107

     7.5.2   Non-metricity   108

     7.5.3   Reconstructing the connection   110

7.6   CURVATURE   111

7.7   TENSOR-VALUED FORMS AND CARTAN'S STRUCTURE EQUATIONS   113

7.8   FORCES AND ACCELERATION   115

     7.8.1   Forces due to scalar fields   115

     7.8.2   Electromagnetic forces   118

7.9   THE FERMI-WALKER DERIVATIVE   120

     7.9.1   Gyroscopes and the concept of rotation   120

     7.9.2   Classical behaviour of a particle with spin   122

CHAPTER   8 ▪ Generalised functions from a geometric perspective   124

8.1   INTRODUCTION   124

8.2   THE EXTERIOR CALCULUS OF LINEAR DISTRIBUTIONS   125

     8.2.1   The relationship with the Dirac delta function   129

8.3   EDGE CURRENTS FROM A GEOMETRICAL PERSPECTIVE   131

8.4   BOUNDARY CONDITIONS IN ELECTROMAGNETISM   140

     8.4.1   Reflection of a pulse from an accelerating mirror   145

APPENDIX   A ▪ Solutions and hints to the exercises   148

A.1   DIFFERENTIAL FORMS   148

A.2   VECTOR FIELDS AND THEIR RELATIONSHIP WITH DIFFERENTIAL FORMS   149

A.3   ASPECTS OF INTEGRATION   150

A.4   THE METRIC TENSOR   152

A.5   MAXWELL'S EQUATIONS IN TERMS OF
       DIFFERENTIAL FORMS   161

A.6   CLASSICAL MECHANICS   164

A.7   CONNECTIONS   171

A.8   GENERALISED FUNCTIONS FROM A GEOMETRIC
       PERSPECTIVE   177

Suggestions for further reading   183

Index   185

Foreword

Undergraduate physics is often taught as a set of disparate fields: classical mechanics here, electrodynamics over there, thermodynamics off to the side, and so forth. Occasionally connections will be revealed, as when statistical mechanics forms a bridge between mechanics and thermodynamics. But in general, the modular structure in which the subject is taught obscures the underlying unity of the discipline.

One particular unifying thread that runs throughout physics is the role played by differential geometry. At first sight, this might seem surprising: differential geometry is often explicitly encountered only in the context of General Relativity, and a student could be forgiven for thinking this is its only use in physics. This could hardly be less true.

Some shadows of geometry are familiar, such as the vector calculus widely used in electromagnetism and fluid dynamics, for instance. But the geometrical nature of these techniques is downplayed, and both the ubiquity and power of the approach are obscured. A good understanding of differential geometry can both highlight the intrinsic nature of certain physical structures and facilitate calculations.

There are many excellent texts on differential geometry. However, in our experience, these tend to emphasise mathematical rigour, which can be daunting to a student first encountering the subject. Our goal with this book is twofold. Firstly, to present a more intuitive approach to the subject, using familiar examples from physics. And secondly, to emphasise how widely geometry stretches throughout physics. In fact, the range of applications of differential geometry in physics is considerably broader than we can discuss here.

We begin with three chapters introducing differential forms and vector fields, the protagonists of our story. These are illustrated with ideas from thermodynamics, Newtonian gravity, magnetostatics and electrostatics. In the fourth chapter, we encounter tensors in general, and the central role (or roles) played by the metric tensor in particular. This brings in elements of relativity theory, cosmology, and fluid dynamics. We then have two chapters that are extended case studies in particular areas of physics (electrodynamics and classical mechanics), emphasising their relationships to particular geometrical structures. Then we conclude with two chapters on more advanced topics: connections and generalised functions, respectively. The seventh and eighth chapters mainly rely on examples from electrodynamics to illustrate the mathematics, although we briefly touch upon Einsteinian gravity. Of course, there are many other advanced topics we could have addressed, most notably Lie groups and their applications in particle physics. However, we feel that the book contains sufficient material to capture the attention of advanced undergraduate and postgraduate physicists.

Each chapter contains numerous exercises, to allow the reader to test their understanding. These range from fairly straightforward to quite demanding. The reader should not be disheartened if they are unable to answer them all; we ask only that they try. Worked solutions are provided in the Appendix.

Differential geometry is a large and growing field, and this book serves only to introduce the basic concepts. We are confident that readers will develop an enthusiasm for the subject, and be keen to learn more. If so, we hope the insights they learn here will equip them well for further study, and make some of the more advanced texts a little less intimidating.

CHAPTER 1

Differential forms

1.1 MOTIVATION

Definite integrals in undergraduate physics are often presented in a manner that emphasises the role of the integration measure. From the outset, the integrand $f(x)$ and the integration measure dx within $\int_a^b f(x)\,dx$ are usually regarded as separate entities. However, this viewpoint does not fully reflect the fact that we can change the integration variable without affecting the result of the integral.

The situation described above is analogous to the invariance of the dot product of a pair of 3-dimensional vectors. Although we can calculate the dot product using Cartesian components of the vectors, the result can be expressed entirely in terms of the lengths of the vectors and the angle between them. The result is independent of the choice of Cartesian frame. Indeed, we do not need a Cartesian frame to picture the dot product; we can simply draw a pair of arrows that emanate from the same point, and label the angle between them. Although Cartesian frames are useful in calculations, they can obscure the physics underpinning the calculation; for example, we do not need a Cartesian frame to draw a picture of the 3-dimensional structure of an electric field. Note that the above analogy was chosen only because the dot product of 3-dimensional vectors is a familiar concept in undergraduate physics; we emphasise that the integral $\int_a^b f(x)\,dx$ need not emerge from a dot product, even when x is a Cartesian coordinate.

Consider the outcome of writing the integral $\int_a^b f(x)\,dx$ in terms of a new integration variable ξ, where the integration range $a \leq x \leq b$ corresponds to the range $\alpha \leq \xi \leq \beta$. Thus

$$\int_a^b f(x)\,dx = \int_\alpha^\beta \phi(\xi)\,\frac{dx}{d\xi}\,d\xi \qquad (1.1)$$

DOI: 10.1201/9781003228943-1

where $\phi(\xi) = f(x)$. The notion of a *differential form* arises when expressing (1.1) in a way that explicitly treats all possible choices of integration variable on an equal footing. The key conceptual step is to combine $f(x)$ and dx into the same object; we introduce the *differential 1-form* $\gamma = f(x)\,dx$ and write the integral as $\int_{\mathcal{I}} \gamma$, where $\mathcal{I}$ is equivalent to the interval $a \leq x \leq b$. The use of the word "equivalent" is important; we do not want to lock ourselves to a particular choice of integration variable. From this perspective, γ is the fundamental mathematical object and the relationship $\gamma = f(x)\,dx$ is a choice that requires extra structure (the variable x). The integration domain $\mathcal{I}$ can be envisaged as a line, and each value of x corresponds to a point on the line. We require that x smoothly increases consistently from one end of the line to the other, but no other conditions on the behaviour of x are necessary. For example, there is no reason why the points in $\mathcal{I}$ corresponding to consecutive integer values of x should be equally spaced. Alternatively, instead of using x, we could choose to express γ and $\mathcal{I}$ as $F(\xi)\,d\xi$ and $\alpha \leq \xi \leq \beta$, respectively. The relationship between $F(\xi)$ and $f(x)$ follows from the fact that they are ingredients in different representations of the same object, γ. We have

$$F(\xi)\,d\xi = f(x)\,dx$$

$$= f(x)\frac{dx}{d\xi}d\xi \tag{1.2}$$

and hence $F(\xi) = f(x)\,dx/d\xi$. Introducing $f(x) = \phi(\xi)$ as we did previously gives $F(\xi) = \phi(\xi)\,dx/d\xi$, and

$$\int_{\mathcal{I}} \gamma = \int_{\alpha}^{\beta} \phi(\xi)\frac{dx}{d\xi}d\xi \tag{1.3}$$

follows immediately, as expected from equation (1.1). Henceforth, for simplicity, we will refer to differential 1-forms as *1-forms*.

Since the sum of a pair of integrals over a given domain can be written as a single integral over the domain, one can immediately see that the sum of a pair of 1-forms is a 1-form. Furthermore, the above can be readily generalised to integrals over curves in spaces whose dimension is greater than one. For example, we can introduce a 1-form $\gamma = f(x,y)\,dx + g(x,y)\,dy$ and calculate the integral $\int_{\mathcal{C}} \gamma$, where the integration domain $\mathcal{C}$ is equivalent to a curve in the xy-plane. Notice that one can view dx and dy as analogous to basis vectors[1] in vector calculus, and regard the functions $f(x,y)$ and $g(x,y)$ as components of γ in this basis.

To proceed further, we need to multiply, and differentiate, 1-forms as well as add them. The next step is to explore the calculus of differential forms, and thermodynamics provides a natural physical context in which to develop the discussion.

1.2 ASPECTS OF THERMODYNAMICS IN TERMS OF DIFFERENTIAL FORMS

The first law of thermodynamics describes the balance of energy of a closed system. It states that the change in internal energy of the system is the sum of the heat gained by the system and the work done on the system during a thermodynamic process. For example, the temperature T, pressure P, volume V, entropy S, and internal energy U of a gas satisfy

$$U_2 - U_1 = \int_1^2 T dS - \int_1^2 P \, dV \qquad (1.4)$$

for any path from point 1 to point 2 in the 2-dimensional space of macroscopic states of the gas. The result of each integral in (1.4) is path-dependent, although the integrals combine to yield a quantity that only depends on the endpoints of the path (the difference between the internal energy U_2 of state 2 and the internal energy U_1 of state 1).

Thermodynamics textbooks typically denote the differentials within each integral in (1.4) as $đQ = T dS$ and $đW = -P \, dV$, respectively, to emphasise the fact that the integrals depend on the details of the path taken. However, in the present context, the symbol $đ$ is superfluous; instead, we will introduce the 1-forms $\mathcal{Q} = TdS$ and $\mathcal{W} = -PdV$ representing heat transfer to the system and work done on the system, respectively. Since $U_2 - U_1 = \int_1^2 dU$, the 1-form

$$dU = \mathcal{Q} + \mathcal{W} \qquad (1.5)$$

follows immediately from (1.4) because the resulting integral holds for all choices of points 1 and 2.

---

[1]It is useful to regard dx and dy as *covectors*, i.e. linear maps on vectors, rather than simply calling them "vectors". Instead of picturing them as arrows, dx and dy should be thought of as elements of surfaces of constant x and constant y, respectively. This subtlety is discussed in detail in Chapter 2.

Exercise 1.1: The equation of state of one mole of an ideal gas is $PV = RT$, where R is the ideal gas constant. Use (1.4) to show that the change in internal energy of the gas due to a change of volume under constant temperature is

$$U_2 - U_1 = T(S_2 - S_1) - RT \ln(V_2/V_1). \qquad (1.6)$$

Differential forms satisfy a set of rules, called *exterior calculus*, that generalises the partial differential calculus of ordinary functions to what might be described as a "calculus of integrands". The rules are given in terms of the *wedge product* (also known as the *exterior product*), denoted $\wedge$, and the *exterior derivative d*. We will introduce their behaviour in the context of the first law of thermodynamics, and then turn to their general properties.

Ordinary functions (such as U, T, S, P, and V) are known as 0-*forms*, and the exterior derivative of any 0-form is a 1-form; for example, dU is a 1-form. The wedge product of any pair of 1-forms is called a 2-*form*; for example, $\mathcal{Q} \wedge \mathcal{W}$ is a 2-form. The wedge product is antisymmetric when its arguments are 1-forms, e.g. $\mathcal{Q} \wedge \mathcal{W} = -\mathcal{W} \wedge \mathcal{Q}$; hence, the wedge product of a 1-form with itself is zero (e.g. $\mathcal{Q} \wedge \mathcal{Q} = 0$). As has already been noted, 1-forms satisfy the same basic rules as vectors; we can multiply them by ordinary functions (0-forms) and we can add the results. Likewise, we can multiply 2-forms by 0-forms and add the results to give another 2-form. Furthermore, the wedge product of a 1-form and a sum of 1-forms can be expanded in the intuitively obvious way, although the ordering of the terms in the wedge product needs to be preserved because of its antisymmetric behaviour (e.g. $\mathcal{Q} \wedge (\mathcal{Q} + \mathcal{W}) = \mathcal{Q} \wedge \mathcal{Q} + \mathcal{Q} \wedge \mathcal{W} = \mathcal{Q} \wedge \mathcal{W}$).

The exterior derivative satisfies a set of rules that resemble ordinary differential calculus, although there are important differences. We have already noted that the exterior derivative of a 0-form is a 1-form. The exterior derivative of a 1-form is a 2-form calculated using the rules

$$d(f\gamma) = df \wedge \gamma + f d\gamma, \qquad (1.7)$$
$$d(dg) = 0 \qquad (1.8)$$

where f and g are 0-forms and γ is a 1-form. For example, the 2-form dQ is

$$dQ = dT \wedge dS + Td(dS)$$
$$= dT \wedge dS \tag{1.9}$$

and, likewise, $dW = -dP \wedge dV$. Since $d(dU) = 0$ using (1.8), the result

$$dT \wedge dS = dP \wedge dV \tag{1.10}$$

follows immediately from (1.5).

The Maxwell relations follow from (1.10) by expressing any pair of the variables T, S, P, and V in terms of the other pair, and making use of the properties of the exterior derivative and wedge product. For example, if T and P are expressed in terms of V and S, then the standard rules of multivariable calculus give

$$dT = \left(\frac{\partial T}{\partial V}\right)_S dV + \left(\frac{\partial T}{\partial S}\right)_V dS, \tag{1.11}$$

$$dP = \left(\frac{\partial P}{\partial V}\right)_S dV + \left(\frac{\partial P}{\partial S}\right)_V dS \tag{1.12}$$

and hence

$$dT \wedge dS = \left(\frac{\partial T}{\partial V}\right)_S dV \wedge dS, \tag{1.13}$$

$$dP \wedge dV = \left(\frac{\partial P}{\partial S}\right)_V dS \wedge dV = -\left(\frac{\partial P}{\partial S}\right)_V dV \wedge dS \tag{1.14}$$

since $dS \wedge dS = 0$, $dV \wedge dV = 0$, and $dS \wedge dV = -dV \wedge dS$. Hence, equations (1.10), (1.13), and (1.14) lead to the Maxwell relation

$$\left(\frac{\partial T}{\partial V}\right)_S = -\left(\frac{\partial P}{\partial S}\right)_V. \tag{1.15}$$

Exercise 1.2: Using the same approach as given above, derive the following Maxwell relations:

$$\left(\frac{\partial T}{\partial P}\right)_S = \left(\frac{\partial V}{\partial S}\right)_P, \quad \left(\frac{\partial S}{\partial V}\right)_T = \left(\frac{\partial P}{\partial T}\right)_V, \quad \left(\frac{\partial S}{\partial P}\right)_T = -\left(\frac{\partial V}{\partial T}\right)_P. \tag{1.16}$$

Exercise 1.3: Use (1.11) and the corresponding expression for dV that arises when V is expressed in terms of T and S to show

$$\left(dT - \left(\frac{\partial T}{\partial V}\right)_S dV\right) \wedge \left(dV - \left(\frac{\partial V}{\partial T}\right)_S dT\right) = 0. \tag{1.17}$$

Hence, by expanding out the brackets in (1.17) and using the rules satisfied by the exterior product, show

$$\left(\frac{\partial T}{\partial V}\right)_S \left(\frac{\partial V}{\partial T}\right)_S = 1. \tag{1.18}$$

Exercise 1.4: Use (1.7), (1.8), and

$$dU = \left(\frac{\partial U}{\partial S}\right)_V dS + \left(\frac{\partial U}{\partial V}\right)_S dV \tag{1.19}$$

to show that the partial derivatives of U with respect to S and V commute.

1.3   THE RULES OF EXTERIOR CALCULUS

It is useful to introduce some new terminology to help explain the general properties of the wedge product and exterior derivative. In particular, the integer p (with $p \geq 0$) in the term p-*form* is called *degree*. For example, the differential forms $\mathcal{Q}$ and $\mathcal{W}$ introduced above have degree 1 (i.e. they are 1-forms). Henceforth, for simplicity, we will refer to "differential forms" as "forms" (in the same spirit as saying "p-form" instead of the more cumbersome "differential p-form").

The calculus of 0-forms, 1-forms, and 2-forms introduced in the previous section generalises to forms of any degree. Let α be a p-form, β be a q-form, γ be a r-form. Any number of forms can be wedged together, and the degree of the result is the sum of the degrees of the terms in the wedge product. For example, the wedge product of the p-form α and q-form β is the $(p+q)$-form $\alpha \wedge \beta$. The wedge product is *associative*, i.e. $(\alpha \wedge \beta) \wedge \gamma = \alpha \wedge (\beta \wedge \gamma)$, and thus we can dispense with the parentheses; instead of writing $(\alpha \wedge \beta) \wedge \gamma$ or $\alpha \wedge (\beta \wedge \gamma)$, we write $\alpha \wedge \beta \wedge \gamma$. Forms of equal degree can be summed to give a form of the same degree; for example, if $r = p$ then $\alpha + \gamma$ is a p-form. Moreover, the wedge product is *distributive*, i.e. we can expand the contents of parentheses in the usual way, although we have to preserve the order of the terms in the wedge product. For example, $(\alpha + \gamma) \wedge \beta = \alpha \wedge \beta + \gamma \wedge \beta$. The order is important because of the symmetry property

$$\alpha \wedge \beta = (-1)^{pq} \beta \wedge \alpha. \tag{1.20}$$

Inspection of (1.20) shows that, under the wedge product, forms of even degree commute with all forms, whilst forms of odd degree anticommute with forms of odd degree. Wedge multiplication by a 0-form is understood by treating the 0-form as an ordinary function; hence, we omit the wedge when multiplying by a 0-form (i.e. we write $f\alpha$ instead of $f \wedge \alpha$ or $\alpha \wedge f$ when f is a 0-form). We will never omit the wedge symbol when multiplying a pair of forms whose degrees are both greater than zero.

The $(p+1)$-form $d\alpha$ results from the action of the exterior derivative d on the p-form α. The exterior derivative satisfies the generalisation

$$d(\alpha \wedge \beta) = d\alpha \wedge \beta + (-1)^p \alpha \wedge d\beta \tag{1.21}$$

of the product rule of differentiation, called the *graded Leibniz rule*, and it satisfies the identity

$$d(d\alpha) = 0. \tag{1.22}$$

Equation (1.22) is commonly written as $d^2\alpha = 0$, and it is a particularly noteworthy property of the exterior derivative because of its significance in topology.

CHAPTER 2

Vector fields and their relationship with differential forms

2.1   INTRODUCTION

So far, we have discussed the exterior algebra of differential forms and explained the relationship between 1-forms and line integrals. In later chapters, we will discuss the general relationship between differential forms and integration. For example, 2-forms correspond to integrands in surface integrals, whilst 3-forms correspond to integrands within volume integrals. In the present chapter, we will focus on the mathematical structures that arise from examining differential forms in the context of linear algebra. However, before turning to the details, it is worth exploring how similar ideas naturally arise from a physical perspective.

The concepts of *potential* and *field* are key ingredients in the description of physical systems. They are often introduced in the physics of classical conservative forces as two sides of the same coin. For example, one can think of the gravitational force on a satellite in orbit around the Earth in terms of a potential or a vector field. The Earth's gravitational potential $\Phi = -GM/r$ (where G is Newton's gravitational constant, M is the mass of the Earth, and r is the distance of the satellite from the Earth's centre) generates the local gravitational acceleration $g = -\nabla\Phi$, a vector field. The vector field g can be represented pictorially in two different ways; as a collection of surfaces or a collection of arrows throughout space. Each surface is an equipotential which, in this case, is a sphere

DOI: 10.1201/9781003228943-2

of radius r. The length and direction of each arrow represents the magnitude and direction of the particular value of g at the base of the arrow. The gravitational force on the satellite is the result of evaluating the product of the mass of the satellite and g at the base of the arrow whose location coincides with the centre of mass of the satellite.

Except for an additive constant, the potential Φ and vector field g contain identical information. However, the relationship $g = -\nabla\Phi$ is deceptively simple; despite its appearance, the calculation of $\nabla\Phi$ requires more than just partial differential calculus because the vector field $\nabla\Phi$ is not the same object as the differential $d\Phi$. The latter can be approximated by the difference of $\Phi(x+\delta x, y+\delta y, z+\delta z)$ and $\Phi(x,y,z)$, where x, y, z are Cartesian coordinates and δx, δy, δz are small displacements in those coordinates. However, from the perspective of vector calculus, the difference of a pair of scalar fields is not a vector field. One more step is required to convert $d\Phi$ into the vector field $\nabla\Phi$; we need to make the replacements $dx \mapsto e_x$, $dy \mapsto e_y$, $dz \mapsto e_z$ in $d\Phi$, where e_x, e_y, e_z are the Cartesian unit vectors of the coordinates x, y, z. This final step may seem natural because, as a consequence, $\nabla\Phi$ points along the path of steepest ascent in Φ but, from the perspective of differential geometry, the details of this replacement require a *metric tensor*. We will leave this concept to a later chapter, but it should now be no surprise that, without a metric tensor, differential 1-forms do not directly correspond to vector fields in vector calculus. In fact, as we will soon see, one should think of 1-forms as linear maps on vector fields (to be more mathematically precise, linear maps on *tangent* vector fields).

2.2 DIFFERENTIAL 1-FORMS AS COTANGENT VECTOR FIELDS

2.2.1 Tangent vector fields

From a physical perspective, the notion of a tangent vector field is a very familiar concept. We have already introduced one example of a tangent vector field; the local gravitational acceleration g. Other tangent vector fields include electric fields and magnetic fields. Any tangent vector field gives rise to a collection of curves (known as *field lines* in physics), and each curve follows the arrows of the tangent vector field. The adjective "tangent" emphasises the fact that the vectors are tangent to curves.

We previously noted that the exterior calculus of differential forms is a "calculus of integrands". Thus, it is perhaps not surprising that a "calculus of differential operators" also plays an important role. In fact, tangent vector fields should be understood as first-order differential operators; instead of $\boldsymbol{g}$, we should consider the operator $\boldsymbol{g} \cdot \boldsymbol{\nabla}$. At first glance, replacing $\boldsymbol{g}$ with its corresponding directional derivative seems like a peculiar strategy that will lead to unnecessarily long-winded expressions. However, the conceptual gain from appreciating that tangent vector fields are actually first-order differential operators is immense. The reason $\boldsymbol{g} \cdot \boldsymbol{\nabla}$ looks strange is simply because we have used the notation of vector calculus. Hence, instead of writing $\boldsymbol{g} \cdot \boldsymbol{\nabla}$, we will introduce the symbol G

$$G = g_x \frac{\partial}{\partial x} + g_y \frac{\partial}{\partial y} + g_z \frac{\partial}{\partial z} \tag{2.1}$$

for the tangent vector field we previously wrote as $\boldsymbol{g} = g_x \boldsymbol{e}_x + g_y \boldsymbol{e}_y + g_z \boldsymbol{e}_z$. Note (2.1) is very natural in the context of field lines because the 0-form

$$Gf = g_x \frac{\partial f}{\partial x} + g_y \frac{\partial f}{\partial y} + g_z \frac{\partial f}{\partial z} \tag{2.2}$$

is the directional derivative of the 0-form f along the field lines of $\boldsymbol{g}$. The functions g_x, g_y, g_z are called the *components* of G in the *basis* with elements $\partial/\partial x$, $\partial/\partial y$, $\partial/\partial z$.[1]

Although G and $\boldsymbol{g}$ are equivalent descriptions of the same physical object, it is important to appreciate that G is a more primitive concept than $\boldsymbol{g}$. The dot product appears in the passage from $\boldsymbol{g}$ to G only because the language of vector calculus is adapted to a particular class of physical calculations. To appreciate this fact, let us now consider the role of tangent vector fields in a physical application where vector calculus does not have a natural use.

2.2.2 Heat capacity from a geometrical perspective

Heat capacities are scalars that encode the heat transferred to a gas corresponding to an infinitesimal change in temperature. Every pair of neighbouring isotherms (curves of constant T) can be connected by paths in the $P - V$ plane and, since every tangent vector field on the $P - V$ plane corresponds to a collection of paths, it is natural to make use of tangent vector fields in this context. Heat capacities are obtained using

---

[1]Henceforth, for linguistic and notational simplicity, we may not explicitly distinguish between a basis (a singular noun) and its elements. For example, we may write "basis" instead of "basis with elements" before a list of elements. However, the precise meaning should be clear from the context.

tangent vector fields that correspond to differentiation with respect to T in the $P - V$ plane.

Every tangent vector field on the $P - V$ plane can be expressed as a linear superposition of the basis $\partial/\partial V$, $\partial/\partial P$ with coefficients (i.e. components) that depend on the coordinates P and V. However, the tangent vector fields of interest here are those that correspond to differentiation with respect to T, and so if X is one such tangent vector field then the components of X must satisfy $XT = 1$. Note that if the 0-form h depends on T only then $Xh = dh/dT$ follows immediately from $XT = 1$ using the chain rule. The remaining freedom in the choice of X is fixed according to the details of the thermodynamic process; in particular, whether the process is undertaken at constant volume or constant pressure. For the rest of this section, we will use the symbol X_V for the tangent vector field corresponding to processes at constant volume and X_P for the tangent vector field corresponding to processes at constant pressure. Thus, the components of X_V are determined by the conditions $X_V T = 1$, $X_V V = 0$, whilst the components of X_P are determined by the conditions $X_P T = 1$, $X_P P = 0$.

Exercise 2.1: Use $X_V V = 0$, $X_V T = 1$ and the general expression

$$X_V = \zeta \frac{\partial}{\partial V} + \xi \frac{\partial}{\partial P} \qquad (2.3)$$

to show

$$X_V = \frac{1}{(\partial T/\partial P)_V} \frac{\partial}{\partial P}. \qquad (2.4)$$

Likewise, show

$$X_P = \frac{1}{(\partial T/\partial V)_P} \frac{\partial}{\partial V}. \qquad (2.5)$$

Heat capacity is commonly defined in thermodynamics textbooks as the path-dependent quantity dQ/dT. However, heat is represented by a 1-form Q from a geometrical perspective, and so we need a procedure for obtaining a 0-form (heat capacity) from a 1-form. We now appeal to the fact that every 1-form can be understood as a linear map on tangent vector fields. This relationship is identical to that between row vectors and column vectors in linear algebra, where a 1-form plays the role of a row vector and a tangent vector field plays the role of a column vector. However, there is one extra vital ingredient; tangent vector fields are also first-order differential operators on 0-forms. The latter property of tangent vector fields is closely related to the relationship between 1-forms and tangent vector fields; to be precise, given any 0-form f, the *contraction* $df(Z)$ of the 1-form df and any tangent vector field Z is defined to have the result

$$df(Z) = Zf. \tag{2.6}$$

All other cases can be obtained by analogy with the behaviour of row vectors and column vectors; for example $(h\,df)(Z) = h\,df(Z) = df(hZ) = hZf$ where h is any 0-form.

Exercise 2.2: Use (2.6) to show that the expressions for X_V, X_P given in (2.4), (2.5) can also be written as the contractions

$$dV(X_V) = 0, \qquad dP(X_V) = \frac{1}{(\partial T/\partial P)_V} \tag{2.7}$$

and

$$dV(X_P) = \frac{1}{(\partial T/\partial V)_P}, \qquad dP(X_P) = 0. \tag{2.8}$$

We now have sufficient geometrical machinery to state the definitions of the heat capacities of a gas at constant volume and pressure. They are

$$C_V = \mathcal{Q}(X_V), \qquad C_P = \mathcal{Q}(X_P) \tag{2.9}$$

where C_V is the heat capacity at constant volume and C_P is the heat capacity at constant pressure.

2.2.3 Heat capacity and the first law of thermodynamics

Relationships between the derivatives of the internal energy U along X_V and X_P and the heat capacities follow from the first law of thermodynamics. Contracting the left-hand side of

$$dU = \mathcal{Q} - P\,dV \tag{2.10}$$

with X_V gives

$$dU(X_V) = X_V U \tag{2.11}$$

whilst the right-hand side yields

$$
\begin{aligned}
(\mathcal{Q} - PdV)(X_V) &= \mathcal{Q}(X_V) - PdV(X_V) \\
&= C_V - PX_V V \\
&= C_V.
\end{aligned}
\tag{2.12}
$$

The first step in (2.12) holds because 1-forms are linear maps on tangent vector fields (c.f. the multiplication of a sum of row vectors and a column vector), whilst the last step holds because X_V is tangent to paths of constant volume in the $P - V$ plane. Thus, the derivative of the internal energy U along X_V is

$$
X_V U = C_V
\tag{2.13}
$$

and, likewise, the derivative of U along X_P,

$$
X_P U = C_P - PX_P V,
\tag{2.14}
$$

follows by contracting (2.10) with X_P.

2.2.3.1 Ideal gas

Equations (2.13), (2.14) readily lead to the well-known relationship

$$
C_P - C_V = nR
\tag{2.15}
$$

between the heat capacities of n moles of ideal gas, where R is the molar gas constant, using only the basic properties of X_P, X_V and ideal gases. Explicit expressions for the components of X_V, X_P are not required.

Since X_P is a first-order linear differential operator, it follows

$$
\begin{aligned}
X_P(PV) &= PX_P V + V X_P P \\
&= PX_P V
\end{aligned}
\tag{2.16}
$$

using the product rule of differentiation and $X_P P = 0$. However,

$$
\begin{aligned}
X_P(nRT) &= nR\, X_P T \\
&= nR
\end{aligned}
\tag{2.17}
$$

since X_P corresponds to first-order differentiation with respect to T. Thus, applying X_P to the ideal gas law

$$
PV = nRT
\tag{2.18}
$$

gives

$$PX_PV = nR. \tag{2.19}$$

However, the internal energy U of an ideal gas can be expressed entirely as a function of temperature T only; thus, $X_VU = dU/dT$, $X_PU = dU/dT$ follow immediately from $X_VT = 1$, $X_PT = 1$, respectively. Thus, the result (2.15) is obtained by subtracting (2.13) from (2.14) and making using of (2.19).

Before ending this section, it is instructive to explore the expressions for X_V, X_P in the two other natural coordinate planes that arise from the thermodynamic variables; the $P - T$ plane and the $T - V$ plane.

Exercise 2.3: Use a similar strategy to that adopted in Exercise 2.1 to derive the expressions

$$X_V = \frac{\partial}{\partial T} - \frac{(\partial V/\partial T)_P}{(\partial V/\partial P)_T}\frac{\partial}{\partial P}, \qquad X_P = \frac{\partial}{\partial T} \tag{2.20}$$

for X_V, X_P in the $P-T$ plane. Hence, using the result for $dV(X_P)$ given in Exercise 2.2, show

$$\frac{1}{(\partial T/\partial V)_P} = \left(\frac{\partial V}{\partial T}\right)_P. \tag{2.21}$$

Likewise, show

$$X_V = \frac{\partial}{\partial T}, \qquad X_P = \frac{\partial}{\partial T} - \frac{(\partial P/\partial T)_V}{(\partial P/\partial V)_T}\frac{\partial}{\partial V} \tag{2.22}$$

hold in the $T - V$ plane, and

$$\frac{1}{(\partial T/\partial P)_V} = \left(\frac{\partial P}{\partial T}\right)_V \tag{2.23}$$

follows as a result.

Exercise 2.3 illustrates an aspect of the geometrical notation for tangent vector fields which sometimes leads to confusion. The symbol $\partial/\partial T$ has been used to denote two different tangent vector fields; one of them is adapted to the $P - T$ plane, whilst the other is adapted to the $T - V$ plane. However, the symbol dT denotes the same 1-form whatever coordinate system is used. The direction of $\partial/\partial T$ at a point is determined by properties of the coordinate line of the non-temperature coordinate (i.e. P or V), as well as the isotherm, at that point, whereas dT has an intrinsic meaning in terms of the exterior derivative d and the 0-form T.

Before drawing this section to a close, we note that 1-forms are also known as *cotangent vector fields* when emphasising their role as linear maps on tangent vector fields. Finally, for convenience and following standard conventions, we will henceforth dispense with the adjective "tangent" and write *vector field* and *covector field* instead of "tangent vector field" and "cotangent vector field", respectively. Thus, covector fields (1-forms) are linear maps on vector fields.

2.3 INTERIOR PRODUCT ON DIFFERENTIAL FORMS

The notion of the contraction of a 1-form and a vector field naturally generalises to an operator on differential forms of arbitrary degree. The *interior product* (or *interior contraction*) of a vector field Z and a p-form α, denoted $i_Z\alpha$, is a $(p-1)$-form calculated as follows. Let f be a 0-form and γ be a 1-form. Then $i_Z f$ is identically zero (there are no differential forms of degree less than zero) and $i_Z\gamma$ is the 0-form given by the contraction of the 1-form γ and Z, i.e. $i_Z\gamma = \gamma(Z)$. The action of i_Z on forms of higher degree follows a similar pattern to the exterior derivative, i.e.

$$i_Z(\alpha \wedge \beta) = i_Z\alpha \wedge \beta + (-1)^p \alpha \wedge i_Z\beta \tag{2.24}$$

where α is a p-form and β is a differential form of any degree. Note $i_Z(f\beta) = f i_Z\beta$ follows immediately from (2.24). Like the contraction of a 1-form and a vector field, the interior product is linear in both of its arguments, i.e. $i_{(Y+Z)}$ is the operator $i_Y + i_Z$, and the action of i_Z on a sum of differential forms is the sum of i_Z acting on each term in the sum. Finally, the repeated application of the same interior product is identically zero, i.e.

$$i_Z i_Z \beta = 0. \tag{2.25}$$

2.3.1 A relationship between thermodynamic properties of a gas

As is the case for the exterior derivative, thermodynamics is an excellent arena in which to demonstrate the use of the interior product. In fact, as we will now show, the well-known relationship

$$C_P - C_V = VT\frac{\beta_P^2}{\kappa_T} \tag{2.26}$$

between the heat capacities, the isobaric expansivity

$$\beta_P = \frac{1}{V}\left(\frac{\partial V}{\partial T}\right)_P, \tag{2.27}$$

and the isothermal compressibility

$$\kappa_T = -\frac{1}{V}\left(\frac{\partial V}{\partial P}\right)_T, \tag{2.28}$$

of a gas follows very naturally using the tools at our disposal.

We begin by noting that the 2-form

$$d\mathcal{Q} = \frac{1}{T}dT \wedge \mathcal{Q} \tag{2.29}$$

is a consequence of the expression $\mathcal{Q} = TdS$ for the heat 1-form $\mathcal{Q}$. Thus

$$i_{X_V}d\mathcal{Q} = \frac{1}{T}i_{X_V}dT \wedge \mathcal{Q} - \frac{1}{T}dT \wedge i_{X_V}\mathcal{Q}$$
$$= \frac{1}{T}X_V T \,\mathcal{Q} - \frac{1}{T}C_V \, dT \tag{2.30}$$

since $i_{X_V}dT = dT(X_V)$, $i_{X_V}\mathcal{Q} = \mathcal{Q}(X_V)$ and the wedge product of a differential form and a 0-form is given by ordinary multiplication. It then follows that the action of i_{X_P} on the 1-form $i_{X_V}d\mathcal{Q}$ gives the 0-form

$$i_{X_P}i_{X_V}d\mathcal{Q} = \frac{1}{T}C_P X_V T - \frac{1}{T}C_V X_P T$$
$$= \frac{C_P - C_V}{T} \tag{2.31}$$

since $X_V T = 1$, $X_P T = 1$. However, as shown earlier, $d\mathcal{Q} = dP \wedge dV$ follows from the first law of thermodynamics, and so $i_{X_P}i_{X_V}d\mathcal{Q}$ can also be written as

$$i_{X_P}i_{X_V}d\mathcal{Q} = X_V P \, X_P V. \tag{2.32}$$

However, expressing the vector fields X_V, X_P in the $P - T$ plane (see Exercise 2.3) leads to the results $X_V P = \beta_P/\kappa_T$, $X_P V = V\beta_P$. Thus, comparing (2.31) with (2.32) immediately reveals (2.26).

2.4   CONSOLIDATION

The presentation so far has been driven entirely by familiar aspects of undergraduate physics. At every stage, we only introduced the mathematical machinery needed for a specific application. It is now time to take stock and, from a general perspective, briefly summarise the properties of all of the mathematical concepts introduced so far and introduce a little more terminology.

2.4.1 Differential forms

Let M be an n-dimensional space with coordinates x^1, x^2, ..., x^n. The examples we have used so far include ordinary 3-dimensional space of everyday experience (where $x^1 = x$, $x^2 = y$, $x^3 = z$ are Cartesian coordinates), and a 2-dimensional space describing the thermodynamic properties of a gas (where, for example, $x^1 = P$, $x^2 = V$). Ordinary functions of the coordinates are known as 0-*forms*, and the differentials dx^1, dx^2, ..., dx^n are known as 1-*forms*. The general 1-form γ on M is a linear combination of dx^1, dx^2, ..., dx^n, i.e.

$$\gamma = \sum_{a=1}^{n} \gamma_a \, dx^a \tag{2.33}$$

where the coefficients γ_1, γ_2, ..., γ_n are 0-forms (i.e. ordinary functions of the coordinates). In particular, the 1-form df is

$$df = \sum_{a=1}^{n} \frac{\partial f}{\partial x^a} \, dx^a \tag{2.34}$$

when expressed in the basis dx^1, dx^2, ..., dx^n. The 0-forms γ_1, γ_2, ..., γ_n are regarded as the components of γ in the basis dx^1, dx^2, ..., dx^n.

One can multiply differential forms using the wedge product $\wedge$. The wedge product of a 0-form f and a 0-form h is the 0-form given by ordinary multiplication (i.e. $f \wedge h = fh$), whilst the wedge product $f \wedge \gamma$ of the 0-form f and the 1-form γ is the 1-form, denoted $f\gamma$, whose components are $f\gamma_1$, $f\gamma_2$, ..., $f\gamma_n$ in the basis dx^1, dx^2, ..., dx^n. The wedge product of p 1-forms is known as a p-*form*, and the integer p is called the *degree* of the p-form. The wedge product is associative and distributive, and it satisfies

$$\alpha \wedge \beta = (-1)^{pq} \beta \wedge \alpha \tag{2.35}$$

where α is a p-form and β is a q-form.

The basis dx^1, dx^2, ..., dx^n for 1-forms gives rise to a basis for p-forms given by all possible linearly independent wedge products of p elements of the basis dx^1, dx^2, ..., dx^n. Since $dx^j \wedge dx^k = -dx^k \wedge dx^j$, it follows that the number of elements of the basis for p-forms is given by the number of distinct ways of choosing p objects from n objects, where the order in which the objects are chosen does not matter. Thus, there are $n!/(p! \, (n-p)!)$ elements of the basis for p-forms in an n-dimensional space. Note that the combinations formula also holds for 0-forms; the sole element of the basis for 0-forms is denoted 1 because $f1 = f$.

2.4.2 The exterior derivative

The 1-form df can be regarded as the result of acting with the operator d, called the *exterior derivative*, that maps p-forms to $(p+1)$-forms. The result of acting with d on differential forms of degree greater than 0 is determined by the rules

$$d(\alpha \wedge \beta) = d\alpha \wedge \beta + (-1)^p \alpha \wedge d\beta, \qquad d(d\beta) = 0, \qquad (2.36)$$

where α is any p-form and β is a differential form of arbitrary degree. Furthermore, the exterior derivative of a sum of q-forms is the sum of the exterior derivative of each q-form. For example, the 2-form $d\gamma$ can be calculated in terms of the partial derivatives of the components γ_1, γ_2, ..., γ_n as follows:

$$\begin{aligned} d\gamma &= \sum_{a=1}^{n} d(\gamma_a dx^a) \\ &= \sum_{a=1}^{n} (d\gamma_a \wedge dx^a + \gamma_a d(dx^a)) \\ &= \sum_{a=1}^{n} \sum_{b=1}^{n} \frac{\partial \gamma_a}{\partial x^b} dx^b \wedge dx^a. \end{aligned} \qquad (2.37)$$

Although the sum in (2.37) is over n^2 terms, only $n!/((n-2)!2!) = n(n-1)/2$ of them are linearly independent because $dx^a \wedge dx^b = -dx^b \wedge dx^a$. Hence, $d\gamma$ can also be written as the linear superposition

$$d\gamma = \sum_{a=1}^{n-1} \sum_{b=a+1}^{n} \left(\frac{\partial \gamma_a}{\partial x^b} - \frac{\partial \gamma_b}{\partial x^a} \right) dx^b \wedge dx^a \qquad (2.38)$$

of linearly independent 2-forms.

Finally, any differential form that is the exterior derivative of another differential form is said to be *exact*. For example, $d\gamma$ is an exact 2-form. A differential form is said to be *closed* if it vanishes under the action of the exterior derivative. Hence, every exact differential form is closed.

2.4.3 Vector fields

Vector fields on M can be thought of as first-order differential operators on 0-forms. In addition to giving rise to the basis dx^1, dx^2, ..., dx^n for 1-forms, the coordinates x^1, x^2, ..., x^n also give rise to the basis

$\partial/\partial x^1, \partial/\partial x^2, \ldots, \partial/\partial x^n$ for vector fields. The general vector field Y on M is

$$Y = \sum_{a=1}^{n} Y^a \frac{\partial}{\partial x^a} \qquad (2.39)$$

where Y^1, Y^2, ..., Y^n are ordinary functions of the coordinates. The coefficients Y^1, Y^2, ..., Y^n are regarded as the components of Y in the basis $\partial/\partial x^1$, $\partial/\partial x^2$, ..., $\partial/\partial x^n$. The action of Y on any 0-form f gives the 0-form Yf where

$$Yf = \sum_{a=1}^{n} Y^a \frac{\partial f}{\partial x^a}. \qquad (2.40)$$

As the notation suggests, any linear combination of vector fields with 0-forms as coefficients gives another vector field; for example, the result of acting with the vector field $X = \lambda Y + \mu Z$ on the 0-form f is

$$Xf = \sum_{a=1}^{n} \left(\lambda Y^a \frac{\partial f}{\partial x^a} + \mu Z^a \frac{\partial f}{\partial x^a} \right) \qquad (2.41)$$

where λ, μ are 0-forms. The positions of λ, μ relative to Y, Z should be carefully noted; the object λY is a vector field, whereas $Y\lambda$ is a 0-form.

In addition to mapping 0-forms to 0-forms via their role as differential operators, vector fields are dual to 1-forms in the sense that 1-forms are linear maps from vector fields to 0-forms. In particular, the *contraction* $\beta(Y)$ of the 1-form β and the vector field Y is the 0-form

$$\beta(Y) = \sum_{a=1}^{n} \beta_a Y^a \qquad (2.42)$$

where

$$\beta = \sum_{a=1}^{n} \beta_a dx^a, \quad Y = \sum_{a=1}^{n} Y^a \frac{\partial}{\partial x^a}. \qquad (2.43)$$

Note that

$$(\lambda \beta + \mu \gamma)(Y) = \lambda \beta(Y) + \mu \gamma(Y) \qquad (2.44)$$

and, likewise,

$$\beta(\lambda Y + \mu Z) = \lambda \beta(Y) + \mu \beta(Z) \qquad (2.45)$$

where β, γ are 1-forms, λ, μ are 0-forms and Y, Z are vector fields. Inspection of (2.42) reveals that the two roles of vector fields are linked through the identity

$$df(Y) = Yf \qquad (2.46)$$

where f is any 0-form. Thus, upon introducing the shorthand $\partial_b = \partial/\partial x^b$, we have the relationship

$$dx^a(\partial_b) = \delta_b^a \tag{2.47}$$

between the bases for 1-forms and vector fields induced from the coordinates $x^1, x^2, \ldots, x^n$, where

$$\delta_b^a = \begin{cases} 1 & \text{if } a = b, \\ 0 & \text{if } a \neq b \end{cases} \tag{2.48}$$

is the Kronecker delta. The bases $\partial_1, \partial_2, \ldots, \partial_n$ and $dx^1, dx^2, \ldots, dx^n$ are said to be *dual*. A basis for vector fields is known as a *frame*, and a basis for 1-forms (i.e. covector fields) is known as a *coframe*.

2.4.4 The interior product

The *interior product* of a vector field Y and a p-form α, denoted $i_Y\alpha$, is the $(p-1)$-form determined by the following rules. The linear operator i_Y annihilates 0-forms (i.e. $i_Y f = 0$ where f is a 0-form) and its action on 1-forms is given by contraction (i.e. $i_Y \gamma = \gamma(Y)$ where γ is a 1-form). Like the exterior derivative, the action of i_Y on differential forms of degree greater than 1 is calculated using the graded Leibniz rule

$$i_Y(\alpha \wedge \beta) = i_Y\alpha \wedge \beta + (-1)^p \alpha \wedge i_Y\beta \tag{2.49}$$

where β is a differential form of any degree. Note that the result of acting repeatedly with i_Y is zero, i.e. $i_Y i_Y \beta = 0$. Finally, as well as being a linear map on differential forms, note that the operator i_Y is linear in its vector argument, i.e. $i_X\beta = \mu i_Y\beta + \lambda i_Z\beta$ where $X = \mu Y + \lambda Z$.

CHAPTER 3

Aspects of integration

3.1 INTRODUCTION

The motivation we gave in Chapter 1 for introducing differential 1-forms was based on considering integrands and integration measures to be two parts of the same entity. For example, we noted that the integral $\int_a^b f(x)\, dx$ can be understood as a particular representation of the integral $\int_{\mathcal{I}} \gamma$. The differential 1-form γ can be expressed as $\gamma = f(x)\, dx$ and the integration domain $\mathcal{I}$ is equivalent to the interval $a \leq x \leq b$. However, physics is replete with integrals over two-dimensional and three-dimensional domains; hence, we need to consider multi-dimensional integration from the perspective of differential forms.

3.2 A UNIFIED PERSPECTIVE ON MULTIVARIABLE INTEGRATION

3.2.1 The emergence of 2-forms in the context of magnetostatics

To proceed, we will now turn to a simple physical example: the magnetic field generated by an electric current flowing steadily through a wire. The magnetic field $\mathbf{H}$ inside the wire satisfies Ampère's law, whose integral version is often expressed as

$$\oint \mathbf{H} \cdot d\mathbf{l} = I \tag{3.1}$$

where I is the total electric current of free charge carriers through any surface enclosed by the Ampèrian path used in (3.1), and $d\mathbf{l}$ is the vector line element of the path. The differential version of Ampère's law is

$$\nabla \times \mathbf{H} = \mathbf{j} \tag{3.2}$$

DOI: 10.1201/9781003228943-3

where $\mathbf{j}$ is the electric current density of the free charge inside the wire. The total electric current $I[\Sigma]$ through any surface Σ inside the wire is

$$I[\Sigma] = \int_\Sigma \mathbf{j} \cdot d\mathbf{S} \qquad (3.3)$$

where $d\mathbf{S}$ is the vector area element of Σ. Thus, (3.1) can be expressed as

$$\int_{\partial\Sigma} \mathbf{H} \cdot d\mathbf{l} = \int_\Sigma \mathbf{j} \cdot d\mathbf{S} \qquad (3.4)$$

where $\partial\Sigma$ is the boundary of Σ, i.e. the closed path around the edge of Σ. The integral version (3.4) and differential version (3.2) of Ampère's law can be immediately obtained from each other using the application

$$\int_\Sigma (\boldsymbol{\nabla} \times \mathbf{H}) \cdot d\mathbf{S} = \int_{\partial\Sigma} \mathbf{H} \cdot d\mathbf{l} \qquad (3.5)$$

of Stokes's theorem. Since Σ can be any surface, it follows that the integrands in the left-hand side of (3.5) and the right-hand side of (3.4) must be equal; thus (3.2) follows as a consequence of (3.4). Conversely, if (3.2) holds then (3.4) follows immediately using (3.5).

As noted previously, integrals over curves can be expressed in terms of 1-forms. In particular, it is natural to introduce the 1-form

$$H = H_x dx + H_y dy + H_z dz, \qquad (3.6)$$

where H_x, H_y, H_z are Cartesian components of $\mathbf{H}$, because

$$\int_{\partial\Sigma} H = \int_{\partial\Sigma} \mathbf{H} \cdot d\mathbf{l} \qquad (3.7)$$

since $d\mathbf{l}$ is the restriction of the differential displacement vector $dx\,\mathbf{e}_x + dy\,\mathbf{e}_y + dz\,\mathbf{e}_z$ to the curve $\partial\Sigma$. However, the natural replacement for $\mathbf{j}$ in the context of exterior calculus is not a 1-form because the integral in (3.4) within which $\mathbf{j}$ resides is over a surface, not a curve. These considerations suggest we should express the right-hand side of (3.4) as

$$\int_\Sigma \mathbf{j} \cdot d\mathbf{S} = \int_\Sigma j \qquad (3.8)$$

where j is a 2-form.

To appreciate why it is possible to replace $\mathbf{j}$ with a 2-form, we begin by observing that a generic 1-form and a generic 2-form both have three independent components in 3-dimensional space. In the present context a generic 1-form is a linear superposition of the 1-form basis elements dx, dy, dz. All 1-forms anticommute with respect to the wedge product, so all 2-forms obtained from the wedge products of dx, dy, dz either vanish or, up to an overall sign, coincide with $dx \wedge dy$, $dy \wedge dz$, or $dz \wedge dx$. Thus, the set $\{dx \wedge dy, dy \wedge dz, dz \wedge dx\}$ is a basis for 2-forms, and we see $\mathbf{j}$ and j have the same number of degrees of freedom.

A natural way to relate the components of $\mathbf{j}$ and the components of the 2-form j is revealed by expressing the left-hand side of (3.2) in terms of the 2-form dH. The exterior derivative of H is readily calculated by considering each term in (3.6). For example,

$$d(H_x dx) = dH_x \wedge dx + H_x d(dx)$$
$$= \frac{\partial H_x}{\partial y} dy \wedge dx + \frac{\partial H_x}{\partial z} dz \wedge dx \qquad (3.9)$$

where the first step is an application of the graded Leibniz rule (1.21), and the second step follows from expanding dH_x as

$$dH_x = \frac{\partial H_x}{\partial x} dx + \frac{\partial H_x}{\partial y} dy + \frac{\partial H_x}{\partial z} dz. \qquad (3.10)$$

Equation (3.9) has been obtained by making use of the properties of the wedge product (in particular, the fact 1-forms anticommute with respect to the wedge product), and the fact the exterior derivative of the exterior derivative of any differential form (in this case, the 0-form x) is zero. Collecting together the three pairs of terms that emerge from expanding the exterior derivative of (3.6) gives

$$dH = j_x dy \wedge dz + j_y dz \wedge dx + j_z dx \wedge dy \qquad (3.11)$$

where

$$j_x = \frac{\partial H_z}{\partial y} - \frac{\partial H_y}{\partial z}, \quad j_y = \frac{\partial H_x}{\partial z} - \frac{\partial H_z}{\partial x}, \quad j_z = \frac{\partial H_y}{\partial x} - \frac{\partial H_x}{\partial y} \qquad (3.12)$$

are the Cartesian components of $\mathbf{j}$. Thus, we identify the electric current 2-form j as

$$j = j_x dy \wedge dz + j_y dz \wedge dx + j_z dx \wedge dy \qquad (3.13)$$

and recast the differential version of Ampère's law, in the context of exterior calculus, as

$$dH = j. \qquad (3.14)$$

Note the identity $dj = 0$ follows immediately from (3.14) due to the fact that the exterior derivative of the exterior derivative of any differential form (in particular, the 1-form H) is zero.

Exercise 3.1: Let k_x, k_y, k_z be the Cartesian components of a vector field $\mathbf{k}$, and introduce the 2-form $k = k_x dy \wedge dz + k_y dz \wedge dx + k_z dx \wedge dy$. Express the exterior derivative of k in terms of the partial derivatives of k_x, k_y, k_z and show that $dk = 0$ is equivalent to $\nabla \cdot \mathbf{k} = 0$.

Before exploring the details of $\int_\Sigma j$, it is worth noting (3.4), (3.7), (3.8), (3.14) together reveal the somewhat striking result

$$\int_\Sigma dH = \int_{\partial\Sigma} H. \tag{3.15}$$

Note the relative positions of the symbols d and ∂. In fact, equation (3.15) is simply Stokes's theorem (3.5) expressed in a more elegant fashion. The importance of this intriguing observation goes beyond mere aesthetics and it will be revisited later in this chapter.

3.2.2 The relationship between integrals of differential forms and ordinary integrals

The integral of a 1-form along a curve is straightforward to reduce to an ordinary 1-dimensional integral. The intuitively obvious result holds true; one simply restricts the 1-form to the curve. For example, the integral $\int_C \alpha$ of the 1-form $\alpha = y\, dx + dy$ along the straight line C starting at $(x = 0, y = 1)$ and ending at $(x = 1, y = 3)$ is $\int_C \alpha = \int_0^1 (2x + 1)dx + \int_0^1 2\, dx = 4$. The final integral is obtained by substituting $y = 2x + 1$ inside the exterior derivative and noting $d(2x + 1) = 2\, dx$.

Exercise 3.2: Show that the integral of the 1-form $\alpha = x\, dy$ around a circle in the xy-plane gives the area of the circle. Likewise, show that the same result is obtained if $\alpha = -y\, dx$.

Integrals of differential forms whose degrees are greater than one are more complicated to handle. For definiteness, let us consider the integral $\int_\Sigma j$ of the electric current 2-form j over the surface Σ. The first step is straightforward; we begin by restricting j to Σ, i.e. replace the coordinates within j with the coordinates of the points in Σ. For example, if Σ is the surface given by $z = f(x, y)$ then

$$
\begin{aligned}
\int_\Sigma j &= \int_\Sigma (j_x dy \wedge dz + j_y dz \wedge dx + j_z dx \wedge dy) \\
&= \int_\Sigma \left(-j_x \frac{\partial f}{\partial x} - j_y \frac{\partial f}{\partial y} + j_z \right) \bigg|_{z=f(x,y)} dx \wedge dy
\end{aligned}
\tag{3.16}
$$

where z has been substituted with f within the exterior derivative (i.e. dz has been replaced by df), and the expression

$$
df = \frac{\partial f}{\partial x} dx + \frac{\partial f}{\partial y} dy
\tag{3.17}
$$

alongside $dy \wedge dx = -dx \wedge dy$ and $dx \wedge dx = dy \wedge dy = 0$ has been used. Except for the presence of the wedge, there is no significant difference between the above procedure and the method we used to evaluate the integral of a 1-form along a curve.

It is now tempting to delete the wedge from the final expression in (3.16) and calculate the integral using standard methods. Although this approach gives the correct result in the present context, greater care is needed in general. We will return to this subtlety shortly; however, for now, we will simply assert

$$
\int_\Sigma j = \int_\mathcal{D} \left(-j_x \frac{\partial f}{\partial x} - j_y \frac{\partial f}{\partial y} + j_z \right) \bigg|_{z=f(x,y)} dx \, dy
\tag{3.18}
$$

i.e.

$$
\int_\Sigma j = \int_\mathcal{D} (\mathbf{j} \cdot \nabla \phi) \big|_{z=f(x,y)} dx \, dy
\tag{3.19}
$$

where $\mathcal{D}$ is the projection of Σ into the $x - y$ plane and the scalar field ϕ is given by $\phi = z - f(x, y)$. Note that the gradient $\nabla \phi$ is calculated *before* setting $z = f(x, y)$, not afterwards, in contrast with the way dz was handled in the final step in (3.16).

The right-hand side of (3.19) is the representation of the integral $\int_\Sigma \mathbf{j} \cdot d\mathbf{S}$ in terms of the level surface $\phi = 0$. Thus, using (3.4), (3.14), (3.15), it is now straightforward to establish the concise integral version

$$
\int_{\partial\Sigma} H = \int_\Sigma j
\tag{3.20}
$$

of Ampère's law.

Exercise 3.3: Consider the parametric representation of Σ given by the position vector

$$\mathbf{r} = x(u, v)\,\mathbf{e}_x + y(u, v)\,\mathbf{e}_y + z(u, v)\,\mathbf{e}_z. \qquad (3.21)$$

By substituting x with $x(u, v)$, y with $y(u, v)$, and z with $z(u, v)$, and subsequently substituting $du \wedge dv$ with $du\,dv$, show

$$\int_\Sigma j = \int_{\mathcal{D}} \mathbf{j} \cdot \left(\frac{\partial \mathbf{r}}{\partial u} \times \frac{\partial \mathbf{r}}{\partial v} \right) du\,dv \qquad (3.22)$$

where $\mathcal{D}$ is the domain of the parameters u, v.

N.B. The right-hand side of (3.22) is simply a parametric representation of the integral $\int_\Sigma \mathbf{j} \cdot d\mathbf{S}$.

The procedure for converting an integral of a differential form to an ordinary multivariable integral requires replacing the differential form with the product of an integrand and an integration element. The approach we took was to simply delete the wedge; for example, we replaced $dx \wedge dy$ with $dx\,dy$ when moving from (3.16) to (3.18). However, it is important to appreciate that this substitution contains an implicit choice of sign because, instead, we could have chosen to replace $dy \wedge dx$ with $dy\,dx$. Note $dy \wedge dx = -dx \wedge dy$ whereas $dx\,dy = dy\,dx$, and so the two different approaches lead to final results that differ by an overall sign. It is clear that, to avoid inconsistencies, care must be taken when deleting wedges. Such considerations are part of the wider concept of *orientation*. However, to avoid unnecessary complication, we will not explore this concept in detail in this chapter. For present purposes, it is enough to assert that Stokes's theorem in vector calculus is reproduced by deleting the wedge from $dx \wedge dy$, $dy \wedge dz$, or $dz \wedge dy$. For consistency, it is important to note the 1-forms in the wedge products in (3.13) are ordered according to the right-hand rule.

Given the above considerations, it is perhaps unsurprising that $dx \wedge dy \wedge dz$ is the appropriate choice of ordering prior to deleting the wedges when integrating a 3-form. To see this, consider the electric displacement $\mathbf{D}$ in a region containing free charges. The integral version of Gauss's law is often expressed as

$$\oint_\Sigma \mathbf{D} \cdot d\mathbf{S} = q \qquad (3.23)$$

where q is the total free charge within the 3-dimensional region enclosed by the Gaussian surface Σ. By analogy with the discussion of differential forms in the context of Ampère's law, it is clear that the role of the vector field $\mathbf{D}$ in (3.23) should be played by the 2-form D given by

$$D = D_x dy \wedge dz + D_y dz \wedge dx + D_z dx \wedge dy. \tag{3.24}$$

Thus, if $\mathcal{U}$ is the 3-dimensional region enclosed by Σ, i.e. $\Sigma = \partial \mathcal{U}$, then we have

$$\int_{\partial \mathcal{U}} D = q. \tag{3.25}$$

Furthermore, recalling Exercise 3.1 earlier in this chapter, the rules of exterior calculus can be used to show

$$dD = \left(\frac{\partial D_x}{\partial x} + \frac{\partial D_y}{\partial y} + \frac{\partial D_z}{\partial z} \right) dx \wedge dy \wedge dz. \tag{3.26}$$

Hence, comparing the results of the application

$$\int_{\mathcal{U}} \nabla \cdot \mathbf{D} \, dx \, dy \, dz = \int_{\partial \mathcal{U}} \mathbf{D} \cdot d\mathbf{S} \tag{3.27}$$

of the divergence theorem with (3.26) leads to

$$\int_{\mathcal{U}} dD = \int_{\partial \mathcal{U}} D. \tag{3.28}$$

As mentioned previously, for consistency with the divergence theorem, we need to order the exterior product of the 1-forms dx, dy, dz as $dx \wedge dy \wedge dz$ prior to deleting the wedges from the left-hand side of (3.28).

3.2.3   The generalised Stokes theorem

It is clear from a comparison of (3.15) with (3.28) that we have encountered a remarkable result. Stokes's theorem and the divergence theorem of vector calculus are simply particular cases of the identity

$$\int_{M} d\omega = \int_{\partial M} \omega \tag{3.29}$$

where ω is a p-form and M is a $(p+1)$-dimensional domain. Equation (3.29) is called the *generalised Stokes theorem*, and it holds for differential forms of *any* degree in spaces of *any* dimension. Even 0-forms are accommodated by (3.29); note ∂M is a pair of points when M is a curve (the start and end of the curve), and $\int_{\partial M} \omega$ is the difference between the values of the 0-form ω at those points.

Before drawing this section to a close, it is worth noting that one can regard ∂ as an operator that acts on domains. It maps a p-dimensional domain M to its $(p-1)$-dimensional boundary ∂M. In particular, the boundary of the boundary of any domain is zero, i.e.

$$\int_{\partial(\partial M)} \omega = 0 \tag{3.30}$$

holds identically for all choices of p-dimensional domain M and $(p-2)$-form ω. Although (3.30) is straightforward to accept based on intuition (e.g. a circle is the boundary of a disc, but the circle has no boundary), it is worth seeing how it immediately follows from the rules of exterior calculus and the generalised Stokes theorem (3.29). Equation (3.29) allows us to write (3.30) as

$$\int_M d(d\omega) = 0 \tag{3.31}$$

which automatically holds regardless of the details of M because the exterior derivative of the exterior derivative of any differential form vanishes. Thus, the fact the boundary of the boundary of any domain and the fact the exterior derivative of the exterior derivative of any differential form both vanish are two sides of the same coin.

3.3   AN ALTERNATIVE PERSPECTIVE ON TWO WELL-KNOWN IDENTITIES IN VECTOR CALCULUS

Before closing this chapter, it is worth discussing two well-known identities in vector calculus that naturally fit into the present context. In particular, recall the curl of the gradient of any scalar field f is zero;

$$\nabla \times (\nabla f) = 0. \tag{3.32}$$

The identity (3.32) is simply $d(df) = 0$ in disguise. Likewise, recall the divergence of the curl of any vector field $\mathbf{A}$ is zero;

$$\nabla \cdot (\nabla \times \mathbf{A}) = 0. \tag{3.33}$$

Perhaps unsurprisingly, the identity (3.33) is another direct consequence of the fact that the exterior derivative of the exterior derivative of any differential form is zero. Thus, from the perspective of the theory of integration of differential forms, (3.32) and (3.33) can be understood as consequences of the fact that the boundary of the boundary of any domain is zero.

Exercise 3.4: Suppose the 1-form α is given in terms of the Cartesian components A_x, A_y, A_z of a vector field $\mathbf{A}$ as $\alpha = A_x dx + A_y dy + A_z dz$. By writing $d(d\alpha)$ explicitly in terms of partial derivatives, show that $d(d\alpha) = 0$ is equivalent to $\nabla \cdot (\nabla \times \mathbf{A}) = 0$.

CHAPTER 4

The metric tensor

4.1 INTRODUCING THE METRIC TENSOR

It hardly needs saying that to make progress in physics, our mathematical framework must contain quantities that allow us to make comparisons with experimental measurements. One such quantity is the *metric tensor*. This object has its origins in the measurement of lengths and angles, but its uses in physics extend far beyond this.

A general tensor is a geometrical object that takes some number of vectors and covectors as input, and outputs a scalar. The action of a tensor is linear in each of its arguments:

$$T(\lambda V + \mu W, X, Y, \cdots) = \lambda T(V, X, Y, \cdots) + \mu T(W, X, Y, \cdots), \quad (4.1)$$

and similarly for the remaining arguments.

The metric tensor g is a rank $(0,2)$ tensor, which means that it acts on two vectors V and W to produce the scalar $g(V,W)$. This is reminiscent of a 2-form; indeed, p-forms are examples of rank $(0,p)$ tensors. However, the defining property of the metric tensor is that it provides the *magnitude* $||V||$ of a vector V:

$$||V|| = \sqrt{|g(V,V)|}. \quad (4.2)$$

Note that the double bars "$|| \cdot ||$" on the left-hand side indicate the magnitude of a vector, while the single bars "$| \cdot |$" on the right-hand side indicate the absolute value of a scalar. If g were a 2-form, the magnitude of any vector would be 0, so the metric tensor must be a different kind of tensor.

A general rank $(0,m)$ tensor can be built up from 1-forms. By definition, a $(0,1)$-tensor *is* a 1-form. Given two 1-forms, α and β, we can

DOI: 10.1201/9781003228943-4

construct the $(0,2)$-tensor $\alpha \otimes \beta$ by its action on two (arbitrary) vectors V and W:

$$[\alpha \otimes \beta](V, W) = \alpha(V)\beta(W), \tag{4.3}$$

where "$\otimes$" is the *tensor product*. We can then construct more general tensors by "tensoring" them together:

$$[T_m \otimes T_n](V_1, \ldots, V_{m+n}) = T_m(V_1, \ldots, V_m)T_n(V_{m+1}, \ldots, V_{m+n}), \tag{4.4}$$

with T_m acting on the first m vectors, and T_n acting on the last n vectors. Hence, the tensor product of a $(0, m)$-tensor with a $(0, n)$-tensor is a $(0, m+n)$-tensor. A $(1, 0)$-tensor is a vector V, and defining its action on a 1-form α as $V(\alpha) = \alpha(V)$, this procedure can be readily generalised to construct tensors of arbitrary rank (m, n).

Finally, we can add together tensors of equal rank,

$$[T + U](V_1, \ldots, V_m) = T(V_1, \ldots, V_m) + U(V_1, \ldots, V_m), \tag{4.5}$$

and multiply them by scalars,

$$[\lambda T](V_1, \ldots, V_m) = \lambda[T(V_1, \ldots, V_m)]. \tag{4.6}$$

The definition (4.3) is reminiscent of the way a 2-form is expressed in terms of a product of two 1-forms, but without the antisymmetry. It follows that we can define the exterior product between two 1-forms as the *antisymmetric part* of their tensor product:

$$\alpha \wedge \beta = \frac{1}{2}(\alpha \otimes \beta - \beta \otimes \alpha). \tag{4.7}$$

In fact, every p-form can be understood as the totally antisymmetric part of a $(0, p)$-tensor.

After this digression into general tensors, we now return to the metric tensor itself. There are a number of properties it must display, in addition to those defining it as a rank (0,2) tensor.

1. **Symmetry**

   As stated previously, the defining property of the metric tensor is its use in providing the magnitude of a vector. Any antisymmetric part of the metric,

$$g_A(V, W) = \frac{1}{2}[g(V, W) - g(W, V)], \tag{4.8}$$

would not contribute to the magnitude. We therefore lose nothing by setting $g_A = 0$ by definition. In other words, the metric tensor is taken to be symmetric,

$$g(V, W) = g(W, V). \tag{4.9}$$

2. **Non-degeneracy**

   When it comes to measuring volumes, we will see that it is important that the metric tensor is *non-degenerate*, i.e. if for all vectors W, $g(V, W) = 0$, then $V = 0$. This property will also be useful in allowing us to use the metric to relate different types of geometric quantity (see Section 4.5).

3. **Positivity?**

   It is sometimes claimed that a true metric tensor must also be positive definite, i.e. for any vector V,

   $$g(V, V) \geq 0, \tag{4.10}$$

   with equality only when $V = 0$. Positive definite metrics are necessarily non-degenerate. With this requirement, our use of absolute values in (4.2) would become redundant.

   There is certainly value in considering positive definite metrics. Our regular notion of "flat[1] Euclidean space" is described by a positive metric, as are many generalisations. But when we want to describe relativistic physics, this is no longer sufficient. The unification of space and time into a single spacetime is achieved with the use of a metric tensor that is not positive definite.

An important characteristic of a metric is its *signature*, which describes how far it deviates from being positive definite. This concept is most clearly understood in the context of *orthonormal frames*, the subject to which we now turn.

---

[1]A proper discussion of curvature, and hence flatness, is deferred to Chapter 7. Here it is sufficient to consider a metric as flat if it *can be written* in the form $g = dx^1 \otimes dx^1 \pm dx^2 \otimes dx^2 \pm \cdots \pm dx^n \otimes dx^n$ where n is the dimension of the space.

4.2 ORTHONORMAL FRAMES

Given coordinates $\{x^a\}$,[2] we can express the metric as

$$g = g_{ab}dx^a \otimes dx^b, \tag{4.11}$$

where the components $\{g_{ab}\}$ are obtained as

$$g_{ab} = g(\partial_a, \partial_b), \tag{4.12}$$

where $\partial_a = \partial/\partial x^a$. Here and in what follows, we have adopted the *Einstein summation convention*, where indices that are repeated, once up and once down, are summed over their range.

Exercise 4.1: Show that (4.12) and (4.11) are mutually consistent.

The symmetry of the metric can be expressed in terms of its components, $g_{ab} = g_{ba}$. Similarly, non-degeneracy can be described as $\det \boldsymbol{g} \neq 0$, where $\boldsymbol{g}$ is the matrix whose entries are $\{g_{ab}\}$. Note that, while $\{g_{ab}\}$, and hence $\det \boldsymbol{g}$, depend on the choice of basis, the requirement $\det \boldsymbol{g} \neq 0$ depends only on the metric g.

The frame $\{\partial_a\}$ is known as a coordinate frame, but there is no requirement to restrict ourselves to such frames. Setting $X_a = M^b{}_a \partial_b$, where the coefficients $\{M^b{}_a\}$ may vary with position, the set $\{X_a\}$ forms a valid frame provided only that the determinant $\det \boldsymbol{M} \neq 0$, where $\boldsymbol{M}$ is the matrix with components $\{M^b{}_a\}$. Associated with the frame $\{X_a\}$ is the dual coframe $\{e^a\}$, defined by $e^a(X_b) = \delta^a_b$, where δ^a_b is the Kronecker delta.

So far, the frame $\{X_a\}$ is arbitrary; the only requirement is that its n elements are linearly independent. For a positive-definite metric, it is always possible to construct a frame such that

$$g = e^1 \otimes e^1 + e^2 \otimes e^2 + \cdots + e^n \otimes e^n, \tag{4.13}$$

i.e. $g_{ab} \doteq \delta_{ab}$, where "$\doteq$" denotes equality in the given frame. Such a frame is known as *orthonormal*: orthogonal ($g(X_a, X_b) = 0$ for $a \neq b$) and normalised ($g(X_a, X_a) = 1$, with no sum over a).

---

[2]More precisely, $\{x^a\}$ is a *set* of coordinates. This abbreviation is similar to that used earlier in the book, where we understood "basis" as "basis with elements". The explicit presence of the set brackets makes it clear that the index a should run over its full range, i.e. $\{x^a\}$ should be understood as $\{x^1, x^2, \ldots, x^n\}$ where n is the dimension of the space.

As an example, consider the metric

$$g = \alpha^2 dx \otimes dx + \beta \left(dx \otimes dy + dy \otimes dx \right) + \gamma^2 dy \otimes dy, \qquad (4.14)$$

in $n = 2$ dimensions, where α, β and γ are functions of (x, y) satisfying $|\beta| < |\alpha\gamma|$. By inspection, the coframe

$$\left\{ e^1 = \alpha dx + \frac{\beta}{\alpha} dy, \quad e^2 = \sqrt{\gamma^2 - \left(\frac{\beta}{\alpha} \right)^2} dy \right\} \qquad (4.15)$$

can be seen to be orthonormal.

Exercise 4.2: Find an alternative orthonormal coframe for this metric, and determine its dual frame $\{X_1, X_2\}$. What goes wrong if $\beta = \alpha\gamma$?

Expanding a pair of vectors in an arbitrary frame, $V = V^a X_a$ and $W = W^a X_a$, the action of the metric on them can be written

$$g(V, W) = g_{ab} V^a W^b. \qquad (4.16)$$

If $\{X_a\}$ is an orthonormal frame, this becomes

$$g(V, W) = V^1 W^1 + V^2 W^2 + \cdots + V^n W^n. \qquad (4.17)$$

If the reader is reminded of the scalar product between vectors, there is good reason for this: the metric tensor is indeed a generalisation of the scalar product from flat Euclidean space to an arbitrary space. The reader may want to check that the traditional "dot product" does indeed satisfy the requirements of a metric.

In what sense, though, is this a *generalisation* of the Euclidean scalar product? If we can express any metric in terms of an orthonormal frame, does it not imply that all metrics are equivalent to the Euclidean scalar product? The answer is both yes and no. At this stage, we need to appreciate a subtlety about where tangent vectors reside.

Just as the tangent to a curve is not part of the curve itself, so tangent vectors do not exist directly on the space M to which they are tangent. Rather, at each point $p \in M$ there is a *vector space*, $T_p M$, which consists of all possible tangent vectors at p.[3] These tangent spaces are

---

[3] Similarly, there is a dual space $T_p^* M$, known as the *cotangent space* at p, which consists of all possible covectors at p.

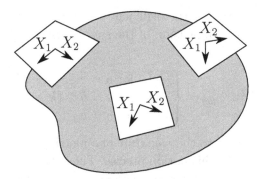

Figure 4.1 The values of an orthonormal frame on the tangent spaces at three different points in a 2-dimensional space.

each equivalent to flat Euclidean space, but tangent spaces at different points, $T_p M$ and $T_q M$, say, with $p \neq q$, are distinct. What makes one metric different from any others is how the tangent spaces are "stitched together", i.e. how an orthonormal frame varies from point to point (see Figure 4.1).

It is important to note that, while an orthonormal frame uniquely specifies a metric tensor, the converse is not true. There is an entire family of orthonormal frames all describing the same metric. Consider a change of coframes:

$$\bar{e}^a = \mathcal{O}^a{}_b e^b. \tag{4.18}$$

The new set of 1-forms $\{\bar{e}^a\}$ is a valid coframe provided only that $\det \mathcal{O} \neq 0$, where $\mathcal{O}$ is the matrix with elements $\{\mathcal{O}^a{}_b\}$. What additional requirements are there for $\{\bar{e}^a\}$ also to be orthonormal? Expanding the metric in the $\{\bar{e}^a\}$ orthonormal coframe,

$$\begin{aligned} g &= \delta_{ab}\, \bar{e}^a \otimes \bar{e}^b \\ &= \mathcal{O}^a{}_c \delta_{ab} \mathcal{O}^b{}_d\, e^c \otimes e^d \\ &= \delta_{cd}\, e^c \otimes e^d \Rightarrow \mathcal{O}^a{}_c \delta_{ab} \mathcal{O}^b{}_d = \delta_{cd}. \end{aligned} \tag{4.19}$$

In terms of matrices, this is $\mathcal{O}^T \mathcal{O} = \mathbb{I}$, where ^{T} denotes transpose and $\mathbb{I}$ is the identity matrix. Hence the transformation matrix $\mathcal{O}$ is a member of the *orthogonal group* $O(n)$ of rotations and reflections in n dimensions.

The above considerations apply to positive metrics, but can be readily generalised to metrics of arbitrary signature. In general, (4.13) is replaced by

$$g = e^1 \otimes e^1 + \cdots + e^k \otimes e^k - e^{k+1} \otimes e^{k+1} - \cdots - e^n \otimes e^n, \tag{4.20}$$

where there are k terms with coefficient $+1$ and $n - k$ terms with coefficient -1, n being the dimension of the space. This can be written more compactly as

$$g = \eta_{ab} e^a \otimes e^b, \quad \eta_{ab} = \begin{cases} +1 & \text{for } a = b \in \{1 \ldots k\}, \\ -1 & \text{for } a = b \in \{k+1 \ldots n\}, \\ 0 & \text{for } a \neq b. \end{cases} \qquad (4.21)$$

We referred earlier to the *signature* of a metric, and we now have the tools we need to make this notion precise. The signature is the difference between the number k of positive terms in the metric (4.20), and the number $n - k$ of negative terms, i.e. $s = 2k - n$.

As with positive metrics, we refer to frames in which the metric takes the form (4.20) as orthonormal. Again, there are many orthonormal frames for a given metric, whose coframes are related by transformations $\bar{e}^a = \mathcal{O}^a{}_b e^b$.

Exercise 4.3: Following the steps leading to (4.19), show in this case that the transformation coefficients satisfy $\mathcal{O}^a{}_c \eta_{ab} \mathcal{O}^b{}_d = \eta_{cd}$. Matrices $\mathcal{O}$ satisfying this relation form a generalisation of the orthogonal group known as $O(k, n - k)$.

While metrics of arbitrary signature are mathematically consistent, there are two signatures that are of particular significance in physics. The first corresponds to positive metrics (also called *Euclidean*), which have signature n. The second corresponds to *Lorentzian* metrics, which have signature $n - 2$ (i.e. only one negative term in (4.20)).[4] For Lorentzian metrics, it is common to denote the basis element attached to the coefficient -1 with the index 0 rather than n.

Exercise 4.4: Consider again the metric

$$g = \alpha^2 dx \otimes dx + \beta \left(dx \otimes dy + dy \otimes dx \right) + \gamma^2 dy \otimes dy \qquad (4.22)$$

of our previous example, but assume now that $|\beta| > |\alpha\gamma|$. By constructing an orthonormal coframe, determine the signature of this metric.

---

[4]Some texts take Lorentzian metrics to have signature $2 - n$, i.e. only one positive term. This does not affect the physics, as other relevant definitions change to compensate this. The $2 - n$ convention is standard in particle physics, while both conventions are in common usage in gravitational physics.

We end this section by noting that coordinate frames and orthonormal frames are not the only options, and there may be others that have significant value. For example, consider (4.22) with $\alpha = \gamma = 0$. While it is possible to find an orthonormal frame, it may be more practical to use the *null frame*

$$n^1 = \sqrt{\beta}dx, \quad n^2 = \sqrt{\beta}dy \tag{4.23}$$

to write $g = n^1 \otimes n^2 + n^2 \otimes n^1$.

4.3 MEASUREMENTS

We began this chapter with the claim that the metric tensor could be used to determine lengths. We have already seen that it can give us the magnitude of a vector, but distances are not in general vectors. In this section, we will explore the role of the metric in calculating lengths of curves, angles between them, and volumes.

4.3.1 Lengths

Consider a curve $\gamma(\lambda)$ on a space M equipped with a positive definite metric g. Given two points $p = \gamma(\lambda_0)$ and $q = \gamma(\lambda_1)$, what is the distance from p to q along γ? An obvious choice might seem to be $\Delta\lambda = \lambda_1 - \lambda_0$. But this cannot be true in general, since the length of the curve is unchanged under reparameterisations ("the distance travelled does not depend on how fast we travel"). Nevertheless, we can find a particular parameter, the change in the value of which does give the distance along the curve. This parameter is known as arc-length, s.

The length of a curve is given by the integral of arc-length along the curve:

$$L = \int_\gamma ds. \tag{4.24}$$

As yet, this is simply the *definition* of arc-length. To understand how to make use of it, we need to find a way to express it in terms of an arbitrary parameter λ.

Note that the tangent vector γ' corresponds to differentiation by λ, and by the chain rule, we can write this in terms of differentiation by s:

$$\gamma' = \frac{\partial}{\partial\lambda} = \frac{ds}{d\lambda}\frac{\partial}{\partial s}. \tag{4.25}$$

By linearity of the metric in each of its arguments, it follows that the 1-form $\sqrt{g(\gamma',\gamma')}d\lambda$ is parameterisation independent. Hence if we take the length of a curve to be given by

$$L = \int_\gamma ||\gamma'||d\lambda, \qquad (4.26)$$

an equivalent definition of arc-length is that it is the unique parameterisation in which the tangent vector is unit normalised:

$$||\gamma'|| = 1. \qquad (4.27)$$

To what extent does the above picture change when we relax the positivity constraint on the metric tensor? A curve may be spacelike, timelike, or null, depending on the properties of its tangent vector. In principle, it is possible for a curve to be, say, spacelike in one region and timelike in another. For the following, we will limit ourselves to segments of a curve which belong to a single category. More generally, a curve can be decomposed into a finite number of such sections, and their path lengths added. However, the physical interpretation in such cases is questionable.

1. A curve is *spacelike* if $g(\gamma',\gamma') > 0$.

   For a spacelike curve, the situation is exactly the same as for a positive definite metric. An infinitesimal segment of a spacelike curve connects different positions at a single moment in time, for some timelike observer. However, in general, it may not be possible to find an observer for which a finite extent of the curve exists at an instant in time.

2. It is *timelike* if $g(\gamma',\gamma') < 0$.

   For a timelike curve, the situation is again similar. The only technical difference is that we need to include the absolute value in the square root when calculating the magnitude of γ', i.e. $||\gamma'|| = \sqrt{-g(\gamma',\gamma')}$. However, the interpretation is subtly different. As suggested by the name, the "length" of such a curve is not really a length at all, but a duration. This duration corresponds to the *proper time* of an object whose four-velocity is γ'. (This assumes units in which the speed of light is unity; in general, the path length is proper time multiplied by c.) A massive particle traces out a timelike curve (its "worldline") in spacetime. Whereas a straight[5] line is (locally) the *shortest* spacelike path between two points, the timelike curve of a non-accelerating worldline is (locally) the *longest* duration between two events. In the "twins paradox", this is what distinguishes the twin who stays on Earth from the one who goes travelling and returns *younger* than their sibling.

3. It is *null* (or *lightlike*) if $g(\gamma', \gamma') = 0$.

   For null curves, the situation is very different. From (4.26), the length of a null curve vanishes. This might suggest we should return to (4.24), and find an alternative way to generalise it. This can indeed be done, but not in a unique way. The only parametrisation-independent length that can be assigned to a null curve is 0.

   How are we to interpret this result? Photons (and other entities that move at the speed of light) follow null paths. Such particles cannot be said to experience the passage of time.

As an example, consider a particle undergoing hyperbolic motion in 2-dimensional Minkowski spacetime, $g_{\text{Mink}} = -dt \otimes dt + dx \otimes dx$. The particle's worldline can be written

$$\gamma(\lambda) = \left\{ t = \lambda, x = \sqrt{\alpha^{-2} + \lambda^2} \right\}, \tag{4.28}$$

where α is the *proper acceleration*. It follows that the tangent vector is

$$\gamma' = \frac{\partial}{\partial t} + \frac{\alpha\lambda}{\sqrt{1 + \alpha^2\lambda^2}} \frac{\partial}{\partial x}, \tag{4.29}$$

with magnitude

$$\sqrt{-g(\gamma', \gamma')} = \frac{1}{\sqrt{1 + \alpha^2\lambda^2}}. \tag{4.30}$$

The proper time elapsed as the particle travels from $\lambda = 0$ to $\lambda = T$ is then

$$\Delta s = \int_0^T \frac{d\lambda}{\sqrt{1 + \alpha^2\lambda^2}} = \frac{1}{\alpha} \sinh^{-1}(\alpha T). \tag{4.31}$$

4.3.2 Angles and rapidities

Suppose now we have two curves, $\gamma_1(\lambda)$ and $\gamma_2(\tau)$,[6] crossing at a point $p = \gamma_1(\lambda_p) = \gamma_2(\tau_p)$. Assuming the space is equipped with a positive

---

[5]For a discussion of "straight lines" and "non-accelerating worldlines", see Chapter 7.

[6]In principle we could parameterise both γ_1 and γ_2 with a single parameter λ, but it is not necessarily convenient to do so.

definite metric, how can we determine the angle between them? In flat Euclidean space, the scalar product of two vectors $\mathbf{t}_1$ and $\mathbf{t}_2$ is given by

$$\mathbf{t}_1 \cdot \mathbf{t}_2 = ||\mathbf{t}_1||\,||\mathbf{t}_2|| \cos \theta. \tag{4.32}$$

We have already noted that the metric tensor is a generalisation of the scalar product. Moreover, the angle between two curves is intrinsically a property of their tangent vectors, so we can simply carry over the definition directly: the angle θ between two curves γ_1 and γ_2 is given by

$$\theta = \cos^{-1}\left(\frac{g(\gamma_1', \gamma_2')}{\sqrt{g(\gamma_1', \gamma_1')g(\gamma_2', \gamma_2')}} \right). \tag{4.33}$$

Exercise 4.5: As with lengths, angles should depend only on the curves, not how they are parameterised. Show that under a reparametrisation $\lambda \to \rho(\lambda)$ of γ_1, the angle θ in (4.33) remains unchanged.

What about curves crossing in a Lorentzian spacetime? We immediately see one problem: for a null curve, $g(\gamma', \gamma') = 0$, so we cannot divide by its square root. When a null curve intersects any other curve, it is simply not possible to assign an angle to their crossing.

If γ_1 and γ_2 are both timelike, we find a different problem: the argument of $\cos^{-1}$ in (4.33) has absolute value greater than 1.[7] If we nevertheless take the formula at face value, it suggests that the angle is *imaginary*. While we shouldn't take this notion too seriously, the cosine of an imaginary angle can be interpreted as the hyperbolic cosine of a real parameter. For crossing timelike curves, therefore, we define the *rapidity*

$$w = \cosh^{-1}\left(\frac{-g(\gamma_1', \gamma_2')}{\sqrt{g(\gamma_1', \gamma_1')g(\gamma_2', \gamma_2')}} \right). \tag{4.34}$$

The rapidity is a measure of how quickly two worldlines are moving relative to one another.

Exercise 4.6: Consider two worldlines in 2D Minkowski space, $g_{\text{Mink}} = -dt \otimes dt + dx \otimes dx$. Taking their tangent vectors to be

$$\gamma_1' = \frac{\partial}{\partial t}, \quad \gamma_2' = \frac{1}{\sqrt{1 - v^2}}\left(\frac{\partial}{\partial t} + v\frac{\partial}{\partial x} \right), \tag{4.35}$$

determine the rapidity w as a function of v.

---

[7]If we used the alternative signature $s = 2 - n$, the argument would be positive, and we would not need the negative sign in (4.34).

If γ_1 is timelike and γ_2 spacelike, or both are spacelike, the argument of the inverse cosine may be less than or greater than 1. However, although we may be able to *calculate* either an angle or a rapidity, its interpretation is rather obscure. Only for spacelike curves which both have vanishing scalar products with a common timelike vector will we use (4.33) to determine an angle.

Nevertheless, we can make use of concepts usually related to angles. In particular, two curves whose scalar products vanish, $g(\gamma_1', \gamma_2') = 0$, will be said to be orthogonal. We will not, though, infer from this that they cross at an angle of $\pi/2$. By this definition, a null curve is orthogonal to itself.

Exercise 4.7: In a Lorentzian spacetime, show that if a timelike curve γ_1 is orthogonal to another curve γ_2, γ_2 must be spacelike.

4.3.3 Volumes

Looking beyond curves, we may also want to know the volume of a region of space. By "volume" here we are referring to the magnitude of an n-dimensional region of an n-dimensional space.

In Chapter 3, we saw that n-forms can be integrated over n-dimensional regions. We now need to decide which n-form to integrate, to give us the volume.

Consider the usual flat Euclidean space with metric

$$g_{\text{Euclid}} = dx \otimes dx + dy \otimes dy. \tag{4.36}$$

From previous experience, we know that volumes (areas) in this space can be obtained by integrating the 2-form $dx \wedge dy$:

$$V_\Sigma = \int_\Sigma dx \wedge dy, \tag{4.37}$$

where Σ is the region whose volume we wish to determine. Why this 2-form? We can gain some insight by switching to polar coordinates, where we have

$$V_\Sigma = \int_\Sigma r dr \wedge d\theta. \tag{4.38}$$

This is unsurprising, since we can easily show $dx \wedge dy = r dr \wedge d\theta$ by interpreting x and y as functions of r and θ in the usual way (i.e. $x =$

$r\cos\theta$ and $y = r\sin\theta$). However, it is telling that both $\{dx, dy\}$ and $\{dr, rd\theta\}$ are orthonormal coframes for g_{Euclid}.

It is natural, therefore, to generalise this by claiming that in an arbitrary space, the volume of a region Σ is given by

$$V_\Sigma = \int_\Sigma e^1 \wedge \cdots \wedge e^n, \tag{4.39}$$

where $\{e^a\}$ is an arbitrary orthonormal coframe. Can we be sure that this will give a unique result? Almost. From the result (4.19), it follows that, if $\{e^a\}$ and $\{\bar{e}^a\}$ are orthonormal coframes, we have

$$\bar{e}^1 \wedge \cdots \wedge \bar{e}^n = \pm e^1 \wedge \cdots \wedge e^n. \tag{4.40}$$

Thus orthonormal coframes fall into two distinct sets, whose volume n-forms differ only in sign. How are we to choose which is the correct volume form? In fact, there is no intrinsic way to decide: we must simply make a choice.

Exercise 4.8: From the definition of "orthonormal", show that two orthonormal coframes $\{e^a\}$ and $\{f^a = \mathcal{O}^a{}_b e^b\}$ in 2 dimensions must satisfy $f^1 \wedge f^2 = \pm e^1 \wedge e^2$.

Suppose we wish to determine the extent of some lower dimensional region, i.e. an m-dimensional region within an n-dimensional space, with $m < n$ (e.g. areas within a 3-dimensional space, or 3-dimensional regions of a 4-dimensional spacetime). We can do this by treating the lower dimensional space as an m-dimensional space in its own right, and determining its m-volume as above. How do we determine the metric on this space?

An m-dimensional subspace of an n-dimensional region can be described by specifying $n - m$ constraints, $\Phi_A = 0$, $A \in \{1, \ldots, n - m\}$. The *induced metric* is obtained by setting $\Phi_A = 0$ and $d\Phi_A = 0$ in the metric of the larger space.

As an example, consider a 2-dimensional paraboloid $\mathcal{P}$ in flat 3-dimensional Euclidean space. It can be described by the constraint $\Phi = z - r^2/(2a)$. Hence we make the substitution $dz = rdr/a$ in $g_{\text{Euclid}} = dr \otimes dr + r^2 d\theta \otimes d\theta + dz \otimes dz$, to obtain the induced metric

$$g_\mathcal{P} = \left(1 + \frac{r^2}{a^2}\right) dr \otimes dr + r^2 d\theta \otimes d\theta. \tag{4.41}$$

Then an orthonormal coframe on $\mathcal{P}$ is $\{e^1 = \sqrt{1+r^2/a^2}dr, e^2 = rd\theta\}$, and the volume form is

$$\Omega = r\sqrt{1+r^2/a^2}dr \wedge d\theta. \tag{4.42}$$

Exercise 4.9: Consider the 2-sphere defined by $\Phi = x^2 + y^2 + z^2 - a^2 = 0$ in 3-dimensional Euclidean space, where a is a constant. Using the induced metric, verify that its volume is $4\pi a^2$.

We briefly return now to the length of a curve. A curve $\gamma(\lambda)$ may be thought of as a restriction of a larger space to a 1-dimensional subspace. The induced metric on this subspace must be $g_\gamma = f(\lambda)d\lambda \otimes d\lambda$, where $f(\lambda)$ is some positive function. Acting with this on the tangent γ' must give the same result as acting with the ambient metric g:

$$g(\gamma',\gamma') = g_\gamma(\gamma',\gamma') = f(\lambda). \tag{4.43}$$

The last equality follows since $\gamma' = \partial/\partial\lambda$. Choosing λ to coincide with arc-length s, it follows that $f(s) = 1$, and hence $g_\gamma = ds \otimes ds$. Clearly $\{ds\}$ is an orthonormal coframe on the subspace (which, being 1-dimensional, has coframes comprising a single element), and we have

$$V_\gamma = \int_\gamma ds. \tag{4.44}$$

Thus the "volume" of a curve agrees with its length, as we would expect.

4.4 EXAMPLES

The above considerations might have felt somewhat abstract. In this section, we will illustrate these ideas by considering in detail a number of metric tensors, on a variety of spaces.

4.4.1 Hyperbolic space

Consider the space of all possible momenta of a particle at a point p in 4D spacetime M. These momenta lie in the cotangent space $T_p^\star M$, which inherits the metric

$$\hat{g} = -dp_0 \otimes dp_0 + dp_1 \otimes dp_1 + dp_2 \otimes dp_2 + dp_3 \otimes dp_3 \tag{4.45}$$

from the metric on M. However, not all points on $T_p^\star M$ are accessible to the particle. Since $-p_0$ is the energy, the particle is constrained to points satisfying the Einstein relation

$$p_0 = -\sqrt{m^2 + (p_1)^2 + (p_2)^2 + (p_3)^2}, \tag{4.46}$$

where m is the particle's mass.

Setting $p_1 = \rho \sin\theta \cos\varphi$, $p_2 = \rho \sin\theta \sin\varphi$, $p_3 = \rho \cos\theta$, it follows that

$$p_0 = -\sqrt{m^2 + \rho^2}, \qquad dp_0 = -\frac{\rho d\rho}{\sqrt{m^2 + \rho^2}}. \tag{4.47}$$

Hence the induced metric on the space H of physically allowed momenta is

$$g = \frac{m^2}{m^2 + \rho^2} d\rho \otimes d\rho + \rho^2 d\theta \otimes d\theta + \rho^2 \sin^2\theta d\varphi \otimes d\varphi. \tag{4.48}$$

As we might have anticipated, in the non-relativistic limit $\rho \ll m$, this reduces to the flat Euclidean metric. In contrast, in the opposite limit $m \to 0$ appropriate for photons, the metric becomes degenerate, i.e. it is no longer a valid metric in this limit.

Exercise 4.10: H can alternatively be described by coordinates $\{u^i\}$, with $p_0 = \sqrt{m^2 + \delta_{ij}u^i u^j}$, $p_i = u^i$ ($i,j \in \{1,2,3\}$). Show that the metric tensor can be written

$$g = \left(\delta_{ij} - \frac{u_i u_j}{m^2 + \delta_{kl}u^k u^l}\right) du^i \otimes du^j, \tag{4.49}$$

where indices on the coordinates u^i are raised and lowered with the Kronecker delta.

4.4.2 Cosmological spacetimes

Much of modern cosmology is based on the *Friedmann-Lemaître-Robertson-Walker (FLRW)* metric,

$$g = -dt \otimes dt + a^2(t)\left[\frac{dr \otimes dr}{1 - kr^2} + r^2(d\theta \otimes d\theta + \sin^2\theta d\varphi \otimes d\varphi)\right]. \tag{4.50}$$

The scale factor $a(t)$ is an increasing function of t. For $k < 0$, notice the similarity with the metric of the hyperbolic space above.

At a given "spatial slice", $t = t_0$, the induced metric is obtained by setting $dt \to 0$ and $a(t) \to a(t_0)$. By inspection, an orthonormal coframe for the induced metric is $\{e^1 = a(t_0)dr/\sqrt{1-kr^2}, e^2 = a(t_0)rd\theta, e^3 = a(t_0)r\sin\theta d\varphi\}$, and so the volume form on such a slice is given by

$$e^1 \wedge e^2 \wedge e^3 = a^3(t_0)\frac{r^2\sin\theta}{\sqrt{1-kr^2}}dr \wedge d\theta \wedge d\varphi. \qquad (4.51)$$

Thus the volume of a region on this slice is proportional to $a^3(t_0)$. Since $a(t)$ is an increasing function, a "fixed" region will increase in volume as t increases. This corresponds to the expansion of the universe.

4.4.3 Rindler spacetime

Consider the *Rindler metric*,

$$g_{\text{Rindler}} = -(1+\alpha X)^2 dT \otimes dT + dX \otimes dX + dY \otimes dY + dZ \otimes dZ, \qquad (4.52)$$

with α a constant. This looks very like the Minkowski metric. In both cases, the induced metric on the "spatial slices" $T = \text{const.}$ are flat Euclidean space. The difference is concentrated entirely in how the part of the metric transverse to these slices varies with X.

An important feature of (4.52) is that it *appears* to become degenerate at $X = -\alpha^{-1}$. The significance of this can be seen by considering a light ray $\gamma(\lambda)$ propagating in the X-direction. The tangent vector to such a light ray can be written

$$\gamma' = \frac{\partial}{\partial T} + (1+\alpha X)\frac{\partial}{\partial X}, \qquad (4.53)$$

where we have parameterised with T (i.e. $T = \lambda$ on $\gamma(\lambda)$), and imposed $g(\gamma', \gamma') = 0$.

With this parameterisation, $1+\alpha X$ gives the rate at which the value of X on the light ray changes as T increases.[8] At $X = -\alpha^{-1}$, this rate vanishes, and the light remains fixed at that position. Hence Rindler spacetime possesses a horizon at $X = -\alpha^{-1}$.

This behaviour appears very different from anything we would expect in Minkowski spacetime. And yet, if we make the coordinate change

$$t = (\alpha^{-1} + X)\sinh\alpha T, \quad x = (\alpha^{-1} + X)\cosh\alpha T, \quad y = Y, \quad z = Z, \qquad (4.54)$$

---

[8]This "coordinate velocity" should not be mistaken for the speed of light, which is determined by the components of γ' in an orthonormal frame, and remains unity for all values of X.

the Rindler metric takes the form

$$g_{\text{Rindler}} = -dt \otimes dt + dx \otimes dx + dy \otimes dy + dz \otimes dz. \qquad (4.55)$$

Thus we see that the Rindler metric is in fact simply the Minkowski metric, written in non-standard coordinates.

This demonstrates an important point: two metrics may appear to be distinct, with very different properties, but can in fact be the same metric, expressed in different coordinates. In general, it is a non-trivial task to demonstrate whether or not two metrics are equivalent.

4.5 DUALITIES

In addition to its role in defining measurable quantities, the metric tensor can be used to convert between different geometric structures. In this section, we explore some of the ways in which this is done.

4.5.1 Metric dual

We have already noted that many physical quantities typically thought of as *vectors* can be better understood as *1-forms*. The reason we can excuse such ambiguities in the nature of physical objects is that the metric tensor allows us to convert freely between vectors and covectors, through a mapping known as the *metric dual*.

Recall that a 1-form α is a linear map acting on vectors: that is, it takes an arbitrary vector V, and turns it into the scalar function $\alpha(V)$. By contrast, the metric tensor is a bilinear map acting on pairs of vectors: it takes arbitrary vectors (V, W) and turns them into the single scalar function $g(V, W)$.

It follows that the action of the metric tensor on a *single*, specific vector V, is itself a 1-form, $g(V, -)$, whose action on an arbitrary vector W is $g(V, -)(W) = g(V, W)$. We call the 1-form $g(V, -)$ the *metric dual* of V.

To avoid unnecessary clutter in our equations, it is convenient to introduce the notation $\tilde{V}$ for the metric dual of a vector V, so that the 1-form $\tilde{V}$ is defined by its action on an arbitrary vector W:

$$\tilde{V}(W) = g(V, W). \qquad (4.56)$$

So far, we have seen how the metric dual converts vectors into 1-forms. In order to complete the equivalence between the two, we must demonstrate the reverse, that to every 1-form α we can associate a unique vector $\tilde{\alpha}$. We define this implicitly in terms of the action of α on an arbitrary vector V:

$$\alpha(V) = g(\tilde{\alpha}, V). \tag{4.57}$$

It is legitimate to question whether this definition *uniquely* defines $\tilde{\alpha}$. If we could find a vector X such that $\tilde{X} = 0$, then $\tilde{\alpha}' = \tilde{\alpha} + X$ would equally satisfy (4.57). However, the requirement that the metric be non-degenerate ensures that $\tilde{X} = 0$ if and only if $X = 0$.

The metric dual allows us to generalise the inner product from an action on pairs of vectors to an action on pairs of 1-forms:

$$G_1(\alpha, \beta) = g(\tilde{\alpha}, \tilde{\beta}). \tag{4.58}$$

G_1 is known as the "inverse metric tensor". It can be expanded in a basis $\{X_a\}$ as

$$G_1 = g^{ab} X_a \otimes X_b, \tag{4.59}$$

where the components g^{ab} form a matrix which is the inverse to that formed by g_{ab}, i.e. $g^{ab} g_{bc} = \delta^a_c$.

Exercise 4.11: In a basis $\{X_a\}$, with dual basis $\{e^a\}$, we can write the metric tensor as $g_{ab} e^a \otimes e^b$. Given a vector $V = V^a X_a$, how would you express $\tilde{V}$ in the dual basis? Given a 1-form $\alpha = \alpha_a e^a$, express $\tilde{\alpha}$ in the basis $\{X_a\}$. Show that $\tilde{\tilde{V}} = V$.

4.5.2 Hodge map

The observant reader will have noticed that, in n dimensions, there are the same number of independent p-forms as there are independent $(n - p)$-forms. This suggests that in some sense there is an equivalence between the two, and once again, the metric tensor allows us to realise this equivalence, through the *Hodge map*.

The Hodge dual $\star$ is a linear map that takes p-forms, and turns them into $(n - p)$-forms. We begin by defining its action on the number 1:

$$\star 1 = e^1 \wedge \cdots \wedge e^n, \tag{4.60}$$

where $\{e^a\}$ is an orthonormal coframe.[9] $\star 1$ is therefore the volume form.

---

[9]As noted above, there are two classes of orthonormal coframe, differing in orientation. In principle, this leads to two Hodge maps, differing only in sign. So long as we are consistent, it does not matter which we choose.

By linearity, this gives the action on all 0-forms:

$$\star f = f \star 1, \tag{4.61}$$

for any 0-form f.

We then define the action of the Hodge map on higher-order forms inductively. For any p-form α and 1-form β, we define

$$\star(\alpha \wedge \beta) = i_{\tilde{\beta}} \star \alpha. \tag{4.62}$$

Again, by linearity this determines the Hodge dual of any p-form.

Exercise 4.12: Show that this definition implies that the Hodge dual of a p-form is indeed an $(n-p)$-form.

Given two 1-forms α and β, the n-form $\alpha \wedge \star\beta$ has a significant role:

$$\alpha \wedge \star\beta = \alpha \wedge i_{\tilde{\beta}} \star 1 = -i_{\tilde{\beta}}(\alpha \wedge \star 1) + (i_{\tilde{\beta}}\alpha) \star 1 = G_1(\alpha, \beta) \star 1, \tag{4.63}$$

where we have used the graded Leibniz rule (2.24) for the interior product, and the fact that the $(n+1)$-form $\alpha \wedge \star 1$ vanishes (since all forms of degree $> n$ vanish identically). This prompts the further generalisation of the metric tensor to an inner product on all p-forms:

$$\gamma \wedge \star\delta = G_p(\gamma, \delta) \star 1, \tag{4.64}$$

where γ and δ are any two p-forms of the same (arbitrary) degree.

Exercise 4.13: If α, β, γ and δ are all 1-forms, determine $G_2(\alpha \wedge \beta, \gamma \wedge \delta)$ in terms of $G_1(\alpha, \gamma)$, $G_1(\alpha, \delta)$, $G_1(\beta, \gamma)$, and $G_1(\beta, \delta)$. Show that the result is symmetric under the exchange $(\alpha, \beta) \leftrightarrow (\gamma, \delta)$.

We have seen that acting twice with the metric dual on a 1-form reproduces the original 1-form (and similarly with vectors). Acting twice with the Hodge dual on a p-form produces a form of degree $n - (n-p) = p$; could this be the original p-form?

Let's explore this question in the simplest case, the 0-form 1:

$$\star\star 1 = \star(e^1 \wedge \cdots \wedge e^n)$$
$$= i_{\tilde{e}^n} \cdots i_{\tilde{e}^1} \star 1$$
$$= G_1(e^1, e^1)G_1(e^2, e^2)\cdots G_1(e^n, e^n). \tag{4.65}$$

This final result is a product of terms each equal to ± 1, so we conclude that $\star\star 1 = \pm 1$. The sign is given by the determinant of the matrix $\boldsymbol{\eta}$ with elements $\{\eta_{ab}\}$, so we write

$$\star\star 1 = (\det\boldsymbol{\eta})1. \tag{4.66}$$

This result generalises to

$$\star\star\alpha = (\det\boldsymbol{\eta})(-1)^{p(n-p)}\alpha \quad \text{for } \alpha \text{ a } p\text{-form}. \tag{4.67}$$

The proof of (4.67) by induction is a good exercise in the manipulation of differential forms, and we encourage the reader to attempt it. A convenient identity for this proof is $\star i_V\alpha = (-1)^{p+1}\tilde{V} \wedge \star\alpha$, where V is a vector and α a p-form. This identity can also be readily proven by induction.

To illustrate the use of the duality properties of the metric tensor, we return to the example of electric current densities. In Chapter 3, we introduced the electric current 2-form $j = j_x dy \wedge dz + j_y dz \wedge dx + j_z dx \wedge dy$, where $\{j_x, j_y, j_z\}$ are the Cartesian components of the electric current vector. Here, we show how this association can be made without appealing to a Cartesian frame, or indeed any frame at all.

In Euclidean space, we have the metric tensor $g_{\text{Euclid}} = dx\otimes dx + dy\otimes dy + dz \otimes dz$. Beginning with the electric current vector $\mathcal{J} = j_x\partial/\partial x + j_y\partial/\partial y + j_z\partial/\partial z$, we first take the metric dual:

$$\tilde{\mathcal{J}} = j_x dx + j_y dy + j_z dz. \tag{4.68}$$

The result is a 1-form, which means we can take its Hodge dual:

$$\star\tilde{\mathcal{J}} = j_x dy \wedge dz + j_y dz \wedge dx + j_z dx \wedge dy. \tag{4.69}$$

Hence we can write $j = \star\tilde{\mathcal{J}}$. Or, regarding the 2-form j as the primitive object, we have

$$\mathcal{J} = \widetilde{\star j}. \tag{4.70}$$

The final result (4.70) makes no reference to Cartesian frames, and indeed can be used with any metric.[10]

Exercise 4.14: Given the electric density 2-form $j = j_\rho d\theta \wedge d\varphi + j_\theta d\varphi \wedge d\rho + j_\varphi d\rho \wedge d\theta$, calculate the electric current vector in the hyperbolic metric (4.48).

---

[10]If we wish to use it on a space with dimension $n \neq 3$, we must take j to be a $(n-1)$-form. This will be important when we consider electrodynamics in Chapter 5.

4.6 CONFORMAL STRUCTURE

In this chapter, we have introduced the metric tensor and explored a number of its applications. Previously we saw that we could make substantial progress without the metric. A natural question, then, is whether there is anything in between: are there structures within the metric we can use, which do not rely on the entire tensor?

Two metric tensors g and g' are *conformally equivalent* if they are related by an overall scaling:

$$g' = \Omega^2 g, \qquad (4.71)$$

where Ω is some positive function of position. The *conformal structure* consists of those properties of a space that are common to all conformally equivalent metrics.

Exercise 4.15: Show that the inner products on 1-forms, G_1 and G_1', are related by $G_1' = \Omega^{-2} G_1$. What is the equivalent relation for the inner products on p-forms, G_p and G_p'?

With the Rindler and Minkowski metrics, we saw that it is not always immediately obvious when two metrics are equal. Similarly, a change of coordinates may be required to identify a conformal equivalence.

Exercise 4.16: The FLRW metric (4.50) with $k = 0$ is conformally equivalent to Minkowski spacetime. Find a function Ω such that $g_{\text{FLRW}} = \Omega^2 g_{\text{Mink}}$. Hint: Introduce a time coordinate τ such that $dt = a(t)d\tau$. You may want to choose a specific form for the scale factor, e.g. $a(t) = At^{2/3}$ (for a matter-dominated universe) or $a(t) = Be^{Ht}$ (for a dark-energy-dominated universe), with A, B and H constants.

A number of physical quantities depend on the metric only through its conformal structure. Here we discuss two examples.

1. **Angles and rapidities**

   One obvious example is the angles between two curves. Under a transformation $g \to \Omega^2 g$, all dependence of θ on Ω in (4.33) cancels out. Similarly, and perhaps less intuitively, the rapidity w in (4.34) depends only on the conformal class of metric, rather than the overall scaling.

2. Causal structure

It is clear that, in general, the magnitude of a vector is not preserved by a transformation $g \rightarrow \Omega^2 g$; rather, $||V|| \rightarrow \Omega ||V||$. In a Lorentzian spacetime, the exception is null vectors for which $||V||=0$. Thus the *causal structure* (i.e. the sign of $g(V,V)$ for any vector V) of spacetime is conformally invariant: the distinction of timelike, spacelike, and null depends only on the conformal structure.

Since the Hodge map is built from the metric, it can change under a scaling transformation. This scaling depends on the degree of the form on which it acts. For a p-form α, we have

$$\star' \alpha = \Omega^{(n-2p)} \star \alpha. \tag{4.72}$$

This can be readily verified by expanding α in an orthonormal basis. (This assumes α itself does not depend on the scaling Ω. This is not necessarily the case: for example, given two vectors V and W, the 0-form $g(V,W)$ clearly depends on Ω if V and W do not.)

It is notable that, in an *even dimensional* space, the Hodge map acting on forms of degree $n/2$ is insensitive to the scale factor. We say that this operator is *conformally invariant*. Given a conformally invariant equation, it is often possible to find an alternative metric that is conformally equivalent to the original, in such a way as to considerably simplify the calculations.

Example: 2D Laplace equation. Many physical quantities are described by the Laplace equation,

$$d \star d\phi = 0, \tag{4.73}$$

where ϕ is a scalar field. Examples include gravity and electrostatics in the absence of sources, and equilibrium heat distributions, among many others. In cases where ϕ is confined to a 2-dimensional surface, or where there is a high degree of symmetry in the third dimension, we can regard this as an equation on a 2-dimensional space.

Consider an example from fluid dynamics. In potential flow, the velocity 1-form is $\nu = d\phi$, where the velocity potential satisfies $d \star d\phi = 0$. Boundary conditions are given by $df \wedge \star d\phi = 0|_{f=0}$ where boundaries are specified by $f = 0$. Suppose we want the flow around an external corner, with boundary $(x = 0, y > 0)$ and $(x > 0, y = 0)$ (see Figure 4.2(a)). The boundary conditions are simpler if we transform to

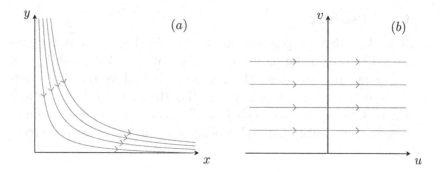

Figure 4.2 Integral curves (stream lines) of the velocity vector field $\tilde{v}$, (a) in the original metric $g_{\text{Euclid}} = dx \otimes dx + dy \otimes dy$, and (b) in the conformally equivalent metric $g' = du \otimes du + dv \otimes dv$.

coordinates $u = (x^2 - y^2)/2$, $v = xy$, where the entire boundary becomes $(v = 0)$ (see Figure 4.2(b)). Then the Euclidean metric takes the form

$$g_{\text{Euclid}} = dx \otimes dx + dy \otimes dy = \frac{du \otimes du + dv \otimes dv}{2\sqrt{u^2 + v^2}}. \qquad (4.74)$$

Since the Laplace equation is conformally invariant, we can work with the alternative metric

$$g' = du \otimes du + dv \otimes dv, \qquad (4.75)$$

where, for example, the velocity potential $\phi = u$ is clearly seen to satisfy both the Laplace equation $d \star d\phi = 0$ and the boundary conditions $dv \wedge \star d\phi = 0|_{v=0}$. In terms of the original coordinates, $\phi = (x^2 - y^2)/2$, giving the velocity 1-form $\nu = xdx - ydy$.

Exercise 4.17: In traditional vector calculus, the Laplace equation is written $\nabla^2\phi = 0$. This is closely related to the Helmholtz equation for time varying electric fields, $\nabla^2\phi + k^2\phi = 0$. Is the Helmholtz equation conformally invariant in 2D? Explain your answer.

4.7 SYMMETRIES AND KILLING VECTORS

In the Rindler and Minkowski metrics, we have seen that it is not always straightforward to decide whether or not two metrics are equivalent. A necessary (but not sufficient) condition for two metrics to be equivalent is that they have the same *symmetries*.

We have an intuitive notion of a symmetry as something remaining fixed under some change. We can formalise this by saying that its derivative vanishes along a "flow",[11] with evolution along that flow representing a transformation.

We know how to differentiate a 0-form along a flow: we act on it with the vector field tangent to the flow. But what about more general tensors? The natural generalisation of the derivative X along a flow is the *Lie derivative*, $\mathcal{L}_X$. The full definition of the Lie derivative is beyond the scope of this book, but here we describe the basic properties that allow us to utilise it.

To use the Lie derivative, we have a few simple rules:

1. On a scalar field f, the Lie derivative is the directional derivative: $\mathcal{L}_X f = Xf$.

2. The Lie derivative commutes with the exterior derivative: $\mathcal{L}_X d\alpha = d\mathcal{L}_X \alpha$.

3. The Lie derivative on tensor products satisfies the Leibniz product rule: $\mathcal{L}_X(T \otimes U) = (\mathcal{L}_X T) \otimes U + T \otimes (\mathcal{L}_X U)$.

4. The Lie derivative commutes with contractions: $\mathcal{L}_X [T(V,-)] = [\mathcal{L}_X T](V,-) + T([\mathcal{L}_X V],-)$.

5. The Lie derivative is linear: $\mathcal{L}_{X+Y} = \mathcal{L}_X + \mathcal{L}_Y$, and $\mathcal{L}_X(T+U) = \mathcal{L}_X T + \mathcal{L}_X U$.

These rules serve to define the Lie derivative on an arbitrary tensor. Some special cases are worth noting:

A. On p-forms, we have *Cartan's identity*: $\mathcal{L}_X = di_X + i_X d$.

This can be proven on 1-forms by expanding both sides on $\alpha = \alpha_a dx^a$. The generalisation to arbitrary p-forms follows by induction, using $\gamma = \alpha \wedge \beta$, with α a 1-form, β a $(p-1)$-form, and using rules 3 and 5. Note that the identity $\mathcal{L}_X(\alpha \wedge \beta) = (\mathcal{L}_X \alpha) \wedge \beta + \alpha \wedge (\mathcal{L}_X \beta)$ follows immediately from rules 3 and 5

---

[11]We can think of a flow as a collection of curves, one passing through each point of the space.

because the wedge product of any collection of 1-forms can be expressed as a linear combination of tensor products of those 1-forms, with constant coefficients. Every differential p-form is the totally antisymmetric part of a $(0, p)$-tensor.

B. On a vector V, $\mathcal{L}_X V = [X, V] \equiv XV - VX$.

This can be proven by acting with $\mathcal{L}_X$ on the contraction $\alpha(V)$, for arbitrary 1-form α. We can obtain the result in two ways: treating $\alpha(V) = \alpha_a V^a$ as a scalar and applying rule 1; and treating $\alpha(V)$ as a contraction and using rules 4 and **A**. Equating the two resulting expressions, all derivatives of α cancel, and we arrive at $\alpha(\mathcal{L}_X V) = \alpha(XV - VX)$. Since α is arbitrary, the desired result is obtained.

Our interest in this section is on the Lie derivatives of metric tensors, for which neither **A** nor **B** apply. Nevertheless, we can readily determine the Lie derivative of any given metric, simply by following the general rules 1–5, as we will now demonstrate.

Symmetries of the metric correspond to vectors K satisfying $\mathcal{L}_K g = 0$. Such vectors are known as *Killing vectors*, and they play an important role in characterising metrics. We end this chapter by illustrating a few examples of Killing vectors.

Consider the flat Euclidean metric in 2 dimensions, $g_{\text{Euclid}} = dx \otimes dx + dy \otimes dy$. We immediately see that the vectors $\partial/\partial x$ and $\partial/\partial y$ are Killing vectors:

$$
\begin{aligned}
\mathcal{L}_{\partial/\partial x} g &= \left(\mathcal{L}_{\partial/\partial x} dx\right) \otimes dx + dx \otimes \left(\mathcal{L}_{\partial/\partial x} dx\right) \\
&\quad + \left(\mathcal{L}_{\partial/\partial x} dy\right) \otimes dy + dy \otimes \left(\mathcal{L}_{\partial/\partial x} dy\right), && \text{(rules 3 \& 5)} \\
&= d\left(\partial x/\partial x\right) \otimes dx + dx \otimes d\left(\partial x/\partial x\right) \\
&\quad + d\left(\partial y/\partial x\right) \otimes dy + dy \otimes d\left(\partial y/\partial x\right), && \text{(rules 2 \& 1)} \\
&= d(1) \otimes dx + dx \otimes d(1) \\
&\quad + d(0) \otimes dy + dy \otimes d(0), && \text{(since } \partial x/\partial x = 1,\ \partial y/\partial x = 0) \\
&= 0, && \text{(since } dC = 0 \text{ for } C \text{ constant)}
\end{aligned}
$$

$$(4.76)$$

and similarly for $\partial/\partial y$. These generate *translations* in the x and y directions, respectively. There is one additional independent Killing vector for this metric, $L = x\partial/\partial y - y\partial/\partial x$:

$$
\begin{aligned}
\mathcal{L}_L g &= d(Lx) \otimes dx + dx \otimes d(Lx) \\
&\quad + d(Ly) \otimes dy + dy \otimes d(Ly), \qquad &\text{(rules 3, 5, 2 \& 1)} \\
&= - dy \otimes dx - dx \otimes dy \\
&\quad + dx \otimes dy + dy \otimes dx, \qquad &\text{(since } Lx = -y \text{ and } Ly = x) \\
&= 0. \qquad &(4.77)
\end{aligned}
$$

L generates *rotations*.

Given the similarities between the Euclidean metric and the 2-dimensional Minkowski metric, $g_{\text{Mink}} = -dt \otimes dt + dx \otimes dx$, one might expect them to have the same Killing vectors (with t substituted for y in the latter). This is true for the translational Killing vectors, but the generator of rotations L does not annihilate g_{Mink}. But g_{Mink} does possess another Killing vector, $K = x\partial/\partial t + t\partial/\partial x$. K generates *boosts*. This distinction between K and L is responsible for the different behaviours of angles and rapidities.

Exercise 4.18: Consider the 2-dimensional Rindler metric $g_{\text{Rindler}} = -(1 + \alpha X)^2 dT \otimes dT + dX \otimes dX$. We know that this is the same as g_{Mink}, so it must possess the same Killing vectors. Verify that $\partial/\partial T$ is a Killing vector, and determine its relationship to the translational and boost Killing vectors discussed above.

CHAPTER 5

Maxwell's equations in terms of differential forms

5.1 INTRODUCTION

It is clear from Chapter 3 that Ampère's law and Gauss's law are readily expressible in terms of differential forms and their integrals. Although any physical theory can be formulated in this language, not every theory can be expressed in such a concise manner. However, electricity and magnetism are particularly well-suited to this formalism. We will now take this discussion a stage further, and explore the general structure of Maxwell's equations from the perspective of exterior calculus.

5.2 THE IMPORTANCE OF THE METRIC

We previously argued that the differential versions of Ampère's law and Gauss's law can both be naturally written as

$$d\omega = \chi \tag{5.1}$$

where ω is p-form and χ is a $(p + 1)$-form. In the case of Ampère's law, ω is a 1-form given by the magnetic field whilst χ is the electric current 2-form. For Gauss's law, ω is the electric displacement 2-form and χ is the electric charge 3-form. Faraday's law follows by letting ω be a 1-form given by the electric field and setting $\chi = 0$. Likewise, Gauss's law for magnetism is given by setting ω to a 2-form obtained from the magnetic induction and setting $\chi = 0$. In this context, without

DOI: 10.1201/9781003228943-5

further comment, we simply replaced any vector field $\mathbf{k}$ of structure $\mathbf{k} = k_x\mathbf{e}_x + k_y\mathbf{e}_y + k_z\mathbf{e}_z$ with the 1-form $k_x dx + k_y dy + k_z dz$ or the 2-form $k_x dy \wedge dz + k_y dz \wedge dx + k_z dx \wedge dy$, depending on which Maxwell equation is under consideration. However, we hinted at the fact the replacements are not arbitrary. In fact, as we will now explain, the procedure we used has a natural formulation in terms of the metric tensor.

Recall that we previously dispensed with the usual way of expressing a vector field, such as $\mathbf{k}$, in favour of the notation

$$K = k_x\frac{\partial}{\partial x} + k_y\frac{\partial}{\partial y} + k_z\frac{\partial}{\partial z}, \tag{5.2}$$

i.e. we replaced $\mathbf{k}$ with the first-order differential operator $K = \mathbf{k} \cdot \boldsymbol{\nabla}$. Furthermore, recall that the metric tensor gives a method for associating vector fields and differential 1-forms in a one-to-one fashion, i.e. every vector field has only one corresponding 1-form, and every 1-form has only one corresponding vector field. The metric tensor of the usual 3-dimensional space of Newtonian physics is[1]

$$\underline{g} = dx \otimes dx + dy \otimes dy + dz \otimes dz. \tag{5.3}$$

The 1-form corresponding to the vector field K is denoted $\widetilde{K}$, and it is given by $\widetilde{K} = \underline{g}(K, -)$. For example, $\widetilde{K} = dz$ if $K = \partial/\partial z$. In general, if K is given by (5.2) then $\widetilde{K} = k_x dx + k_y dy + k_z dz$, which is the 1-form replacement for $\mathbf{k}$.

The 2-form replacement for $\mathbf{k}$ requires the metric $\underline{g}$ and the Hodge map $\underline{\star}$ to construct. The volume 3-form $\underline{\star}1$ is given by

$$\underline{\star}1 = dx \wedge dy \wedge dz \tag{5.4}$$

and $\underline{\star}\widetilde{K}$ is the required 2-form. This conclusion immediately follows by using the identity $\underline{\star}\widetilde{K} = i_K\underline{\star}1$, since

$$i_K(dx \wedge dy \wedge dz) = i_K dx \wedge dy \wedge dz - dx \wedge i_K dy \wedge dz + dx \wedge dy \wedge i_K dz$$
$$= k_x\, dy \wedge dz - k_y\, dx \wedge dz + k_z\, dx \wedge dy \tag{5.5}$$

follows from the rules obeyed by the interior product and $i_K dx = dx(K) = k_x$, etc. The desired result

$$\underline{\star}\widetilde{K} = k_x dy \wedge dz + k_y dz \wedge dx + k_z dx \wedge dy \tag{5.6}$$

is obtained immediately because $dx \wedge dz = -dz \wedge dx$.

---

[1]In this chapter, to avoid confusion, we will reserve the symbol g for the space-time metric and $\star$ for the Hodge map on forms in spacetime. The symbols $\underline{g}$, $\underline{\star}$ are associated with 3-dimensional space.

The above considerations immediately lead to the conclusion that Gauss's law $\nabla \cdot \mathbf{D} = \rho$ can be written in terms of the vector field $\mathcal{D}$,

$$\mathcal{D} = D_x \frac{\partial}{\partial x} + D_y \frac{\partial}{\partial y} + D_z \frac{\partial}{\partial z}, \tag{5.7}$$

as

$$d_{\star}\widetilde{\mathcal{D}} = \rho_{\star}1 \tag{5.8}$$

where ρ is the charge density, and D_x, D_y, D_z are the Cartesian components of the electric displacement $\mathbf{D}$.

Exercise 5.1: Show that Ampère's law $\nabla \times \mathbf{H} = \mathbf{j}$ can be written as

$$d\widetilde{\mathcal{H}} = \star \widetilde{\mathcal{J}} \tag{5.9}$$

where

$$\mathcal{H} = H_x \frac{\partial}{\partial x} + H_y \frac{\partial}{\partial y} + H_z \frac{\partial}{\partial z}, \qquad \mathcal{J} = j_x \frac{\partial}{\partial x} + j_y \frac{\partial}{\partial y} + j_z \frac{\partial}{\partial z} \tag{5.10}$$

correspond to the magnetic field $\mathbf{H}$ and current density $\mathbf{j}$, respectively.

5.3 THE VACUUM MAXWELL EQUATIONS FROM A FOUR-DIMENSIONAL PERSPECTIVE

The previous considerations are valid only for static electric fields and static magnetic fields. Whilst there are advantages in formulating electrostatics and magnetostatics using exterior calculus, the most significant gains are obtained when the fields are time dependent. We can exploit the profound physical fact that electric and magnetic fields are merely different parts of the same mathematical object: a 2-form in spacetime. As a result, the theory of electromagnetism is remarkably concise when it is written in terms of differential forms in spacetime.

Let us begin by focussing on the vacuum Maxwell equations, i.e. the equations describing the behaviour of electric and magnetic fields in a region of space that is otherwise empty. For simplicity, throughout the remainder of this chapter, we will use units in which the permittivity ϵ_0 and permeability of the vacuum μ_0 are equal to unity (and, hence, $c = 1$ in those units); hence $\mathbf{D} = \mathbf{E}$ and $\mathbf{H} = \mathbf{B}$. Thus, Gauss's law for magnetism, Faraday's law of induction, Gauss's law, and the Ampère-Maxwell law, are

$$\mathbf{\nabla} \cdot \mathbf{B} = 0, \qquad \mathbf{\nabla} \times \mathbf{E} = -\frac{\partial \mathbf{B}}{\partial t}, \tag{5.11}$$

$$\mathbf{\nabla} \cdot \mathbf{E} = 0, \qquad \mathbf{\nabla} \times \mathbf{B} = \frac{\partial \mathbf{E}}{\partial t}, \tag{5.12}$$

respectively.

The above formulation of Maxwell's equations is not adapted to the fact that their structure is completely independent of the inertial frame in which they are expressed. To be more precise, the relationship between the pair of fields $\mathbf{E}'$, $\mathbf{B}'$ measured by an observer moving at constant velocity with respect to a stationary observer, and the pair of fields $\mathbf{E}$, $\mathbf{B}$ measured by the stationary observer, is such that

$$\mathbf{\nabla}' \cdot \mathbf{B}' = 0, \qquad \mathbf{\nabla}' \times \mathbf{E}' = -\frac{\partial \mathbf{B}'}{\partial t'}, \tag{5.13}$$

$$\mathbf{\nabla}' \cdot \mathbf{E}' = 0, \qquad \mathbf{\nabla}' \times \mathbf{B}' = \frac{\partial \mathbf{E}'}{\partial t'}, \tag{5.14}$$

where $\mathbf{\nabla}'$ is the del operator and t' is the time coordinate in the inertial frame of reference of the moving observer. In fact, the relationship between the fields is such that the 2-form F given by

$$\begin{aligned} F = E_x \, dt \wedge dx + E_y \, dt \wedge dy + E_z \, dt \wedge dz \\ - B_x \, dy \wedge dz - B_y \, dz \wedge dx - B_z \, dx \wedge dy \end{aligned} \tag{5.15}$$

has the same structure regardless of the frame used, i.e.

$$\begin{aligned} F = E'_{x'} \, dt' \wedge dx' + E'_{y'} \, dt' \wedge dy' + E'_{z'} \, dt' \wedge dz' \\ - B'_{x'} \, dy' \wedge dz' - B'_{y'} \, dz' \wedge dx' - B'_{z'} \, dx' \wedge dy'. \end{aligned} \tag{5.16}$$

In other words, the components of the electric field and magnetic field are different components of a 2-form on spacetime (variously known as the *electromagnetic*, *Maxwell* or *Faraday* 2-form), and the relationship between the fields in different frames of reference can be deduced using the appropriate coordinate transformation. For example, suppose that the moving observer has constant speed v along the x-axis. The Lorentz transformation between the coordinates of the stationary observer and the moving observer is

$$t' = \frac{t - vx}{\sqrt{1 - v^2}}, \quad x' = \frac{x - vt}{\sqrt{1 - v^2}}, \quad y' = y, \quad z' = z. \tag{5.17}$$

Furthermore, suppose that the electric field in the moving observer's frame is parallel to the y'-axis, and the magnetic field in their frame is zero. Hence, the electromagnetic 2-form is

$$F = E'_{y'} \, dt' \wedge dy'. \tag{5.18}$$

Plugging the expressions

$$dt' = \frac{dt - v \, dx}{\sqrt{1 - v^2}}, \quad dy' = dy \tag{5.19}$$

into (5.18) gives

$$F = \frac{E'_{y'}}{\sqrt{1 - v^2}} dt \wedge dy - \frac{v E'_{y'}}{\sqrt{1 - v^2}} dx \wedge dy \tag{5.20}$$

which, when compared to (5.15), reveals

$$E_y = \frac{E'_{y'}}{\sqrt{1 - v^2}}, \quad B_z = \frac{v E'_{y'}}{\sqrt{1 - v^2}} \tag{5.21}$$

with the other components of **E** and **B** being zero.

The field equations satisfied by F corresponding to (5.11), (5.12) are remarkably simple. Gauss's law for magnetism and Faraday's law of induction both emerge from the same equation,

$$dF = 0, \tag{5.22}$$

known as the *Gauss-Faraday law*.

Exercise 5.2: Use the properties of the exterior derivative and the wedge product to show

$$d(E_x dt \wedge dx) = \partial_y E_x dt \wedge dx \wedge dy - \partial_z E_x dt \wedge dz \wedge dx \tag{5.23}$$

and

$$d(B_x dy \wedge dz) = \partial_t B_x dt \wedge dy \wedge dz + \partial_x B_x dx \wedge dy \wedge dz \tag{5.24}$$

where $\partial_t, \partial_x, \partial_y, \partial_z$ are shorthand for $\partial/\partial t, \partial/\partial x, \partial/\partial y, \partial/\partial z$, respectively.

Obtain the corresponding results for $d(E_y dt \wedge dy)$, $d(E_z dt \wedge dz)$, $d(B_y dz \wedge dx)$, and $d(B_z dx \wedge dy)$, by cyclically permuting x, y, z or using direct computation. Hence, show (5.22) is equivalent to (5.11).

Since the exterior derivative of the exterior derivative of any differential form is zero, we immediately obtain the solution $F = dA$ to (5.22), where A is a 1-form potential. Note that the electromagnetic 2-form F is invariant with respect to the gauge transformation $A \mapsto A + d\lambda$, where the gauge function λ is any 0-form. The introduction of the 1-form A is equivalent to introducing the expressions $\mathbf{E} = -\boldsymbol{\nabla}\varphi - \partial_t\mathbf{A}$, $\mathbf{B} = \boldsymbol{\nabla} \times \mathbf{A}$ for the fields $\mathbf{E}$, $\mathbf{B}$ in terms of the potentials φ, $\mathbf{A}$. However, the present approach is clearly much more succinct.

The spacetime metric g and its Hodge map are required to unify the remaining two Maxwell equations (Gauss's law and the Ampère-Maxwell law). The metric of Minkowski spacetime is

$$g = -dt \otimes dt + dx \otimes dx + dy \otimes dy + dz \otimes dz \qquad (5.25)$$

and its volume 4-form is

$$\star 1 = dt \wedge dx \wedge dy \wedge dz. \qquad (5.26)$$

Care must be taken to properly account for the different signs in (5.25). For example, $\star dt = -\star \tilde{\partial}_t = -i_{\partial_t} \star 1 = -dx \wedge dy \wedge dz$, whereas $\star dy = dt \wedge dx \wedge dz$.

Exercise 5.3: Use the rule $\star(\alpha \wedge \tilde{X}) = i_X \star \alpha$, where α is any differential form and X is any vector field, to show

$$\star(dt \wedge dx) = -dy \wedge dz, \qquad \star(dy \wedge dz) = dt \wedge dx, \qquad (5.27)$$
$$\star(dt \wedge dy) = dx \wedge dz, \qquad \star(dx \wedge dz) = -dt \wedge dy, \qquad (5.28)$$
$$\star(dt \wedge dz) = -dx \wedge dy, \qquad \star(dx \wedge dy) = dt \wedge dz. \qquad (5.29)$$

Hence, show $\star\star\beta = -\beta$ where β is any 2-form. Check this agrees with the general result (4.67).

The *Gauss-Ampère law*

$$d \star F = 0 \qquad (5.30)$$

unifies Gauss's law and the Ampère-Maxwell law. It is the counterpart to the Gauss-Faraday law (5.22).

Exercise 5.4: Use the results of Exercise 5.3 and (5.15) to show

$$\star F = -\,B_x\,dt \wedge dx - B_y\,dt \wedge dy - B_z\,dt \wedge dz$$
$$- E_x\,dy \wedge dz - E_y\,dz \wedge dx - E_z\,dx \wedge dy. \qquad (5.31)$$

Hence, show (5.30) is equivalent to (5.12).

Before moving on to a specific application of the above formalism, it is worth noting that the Hodge map $\star$ and electromagnetic 2-form F naturally yield a pair of 0-forms given by $X = \star(F \wedge \star F)$, $Y = \star(F \wedge F)$ called *electromagnetic invariants*.

Exercise 5.5: Use (5.15) and (5.31) to obtain the expressions

$$X = |\mathbf{E}|^2 - |\mathbf{B}|^2, \qquad Y = 2\mathbf{E} \cdot \mathbf{B} \qquad (5.32)$$

for the electromagnetic invariants.

5.4 ELECTROMAGNETIC WAVES FROM A SPACETIME PERSPECTIVE

The advantages of using the Gauss-Faraday and Gauss-Ampère laws directly, in preference to (5.11) and (5.12), are computational as well as conceptual. For example, consider a linearly polarised electromagnetic plane wave propagating along the x-axis, whose polarisation vector is along the y-axis. Its electromagnetic 2-form is

$$F = f(u)du \wedge dy \qquad (5.33)$$

where $u = x - t$ is a null coordinate and $f(u)$ is any function of u. Since f only depends on u, it follows $df \wedge du = 0$ and therefore $d(f du) = 0$ because the exterior derivative of an exterior derivative is zero. Thus, $dF = 0$ as required. Furthermore, it is straightforward to show $\star(du \wedge dy) = -du \wedge dz$ using the results of Exercise 5.3. Hence $\star F = -f du \wedge dz$ and the result $d \star F = 0$ follows immediately using the same approach.

It is also worth noting that the results $F \wedge F = 0$ and $F \wedge \star F = 0$ follow immediately because $du \wedge du = 0$; hence, the electromagnetic invariants both vanish in this case.

The simplicity of the previous calculation is worth comparing to the details of the equivalent calculation using (5.11), (5.12).

Exercise 5.6: Let $\mathbf{e}_x$ be a unit vector in the x direction, $\mathbf{e}_y$ be a unit vector in the y direction, and $\mathbf{e}_z$ be a unit vector in the z direction. Show that (5.33) is equivalent to $\mathbf{E} = -f(x - t)\,\mathbf{e}_y$, $\mathbf{B} = -f(x - t)\,\mathbf{e}_z$. Substitute $\mathbf{E}$, $\mathbf{B}$ in (5.11), (5.12) and show they are satisfied.

Further insight into the above perspective on electromagnetic plane waves readily emerges from an examination of the general properties of the Hodge map. In Exercise 5.3 we showed $\star \star \beta = -\beta$ where β is any 2-form. Thus, upon introducing the imaginary unit[2] $i = \sqrt{-1}$, we see that the linear operator $i\star$ satisfies $(i\star)^2 \beta = \beta$. Moreover, $\star\beta$ is itself a 2-form and so any linear combination of β and $\star\beta$ is also a 2-form. These observations suggest the introduction of the pair of linear operators

$$P = \frac{1}{2}(1 - i\star), \qquad \bar{P} = \frac{1}{2}(1 + i\star) \tag{5.34}$$

on the space of 2-forms. Inspection of the properties of P, $\bar{P}$ shows that they are projection operators and they decompose the set of 2-forms into two disjoint subsets of 2-forms. Every element of each subset is an eigenvector of the Hodge map.[3]

Exercise 5.7: Show that P, $\bar{P}$ satisfy $P + \bar{P} = 1$, $PP = P$, $\bar{P}\bar{P} = \bar{P}$ and $P\bar{P} = \bar{P}P = 0$ when acting on 2-forms.

It is important to appreciate that the components with respect to dt, dx, dy, dz of any 2-form in either of the two subsets must be complex because i is present in (5.34).

The above decomposition is compatible with the vacuum Maxwell equations (5.22), (5.30) in the sense that any closed element of either

---

[2] The symbol i used here has nothing to do with the interior product on differential forms.

[3] It is not uncommon to describe each element as being *self-dual* or *anti-self-dual*, although care must be taken as the duality is with respect to the operator $i\star$, not the Hodge map $\star$. We will not use this terminology here to avoid confusion.

subset solves them. A closed differential form is a form whose exterior derivative is zero. To show that the vacuum Maxwell equations are satisfied, we will focus on the subset of 2-forms preserved by P (i.e. the subset annihilated by $\bar{P}$), although it is straightforward to repeat the exercise using the other subset. Let $\mathbb{F}$ be any 2-form (with complex components) that satisfies

$$d\mathbb{F} = 0, \qquad P\mathbb{F} = \mathbb{F}. \qquad (5.35)$$

It can then be shown that (5.22), (5.30) are solved by $F = \mathrm{Re}(\mathbb{F})$, as follows.[4] The Gauss-Faraday law (5.22) is obtained using $\mathrm{Re}(\mathbb{F}) = (\mathbb{F} + \bar{\mathbb{F}})/2$ and $d\bar{\mathbb{F}} = \overline{d\mathbb{F}} = 0$, where the bar denotes complex conjugation. Likewise, the Gauss-Ampère law (5.30) is satisfied because $\star F = i(\mathbb{F} - \bar{\mathbb{F}})/2$ follows from $\star \mathbb{F} = i\mathbb{F}$ (since $P\mathbb{F} = \mathbb{F}$) and $\overline{\star \mathbb{F}} = \overline{i\mathbb{F}} = -i\bar{\mathbb{F}}$. It is straightforward to repeat the above argument if $\mathbb{F}$ is chosen to satisfy $\bar{P}\mathbb{F} = \mathbb{F}$ instead of $P\mathbb{F} = \mathbb{F}$.

One of the steps in the previous argument made use of the result $\overline{\star \mathbb{F}} = \star \bar{\mathbb{F}}$. Of course one can verify this result directly by expressing $\mathbb{F}$ using the basis dt, dx, dy, dz, but it is clear that the result must hold because the Hodge map just swaps components of differential forms. The details of those components are unaffected during their transit; in particular, whether they are real or complex is irrelevant.

The above discussion shows that the closed 2-forms preserved by either P or $\bar{P}$ are intimately tied to the solutions of the vacuum Maxwell equations. In particular, consider the 2-form

$$\mathbb{F} = \mathfrak{f}(u, \zeta) \, du \wedge d\zeta \qquad (5.36)$$

where $\zeta = y + iz$ and $\mathfrak{f}(u, \zeta)$ is a complex function of $u = x - t$ and ζ only. Clearly, $d\mathbb{F} = 0$ follows immediately because $d\mathfrak{f} = \partial_u \mathfrak{f} \, du + \partial_\zeta \mathfrak{f} \, d\zeta$ and thus $d\mathfrak{f} \wedge du \wedge d\zeta = 0$. Furthermore, $P\mathbb{F} = \mathbb{F}$ follows because $\star(du \wedge d\zeta) = i \, du \wedge d\zeta$ using (5.27), (5.28), (5.29). The linearly polarised electromagnetic plane wave (5.33) is immediately obtained from $F = \mathrm{Re}(\mathbb{F})$ when $\mathfrak{f}(u, \zeta) = f(u)$. Choices of $\mathfrak{f}$ that are non-constant with respect to ζ describe electromagnetic waves inside physical structures that permit TEM (Transverse Electro-Magnetic) modes of propagation. The electric field and magnetic field of any TEM mode are orthogonal, which accords with (5.36) because $du \wedge du = 0$, and therefore $F \wedge F = 0$

---

[4]It is worth noting that we could have chosen $F = \mathrm{Im}(\mathbb{F})$ instead of $F = \mathrm{Re}(\mathbb{F})$. This flexibility corresponds to the fact that the vacuum Maxwell equations are invariant under the replacement $F \mapsto \star F$ (so $\star F \mapsto \star \star F = -F$), a property known as *electromagnetic duality*.

since $F = du \wedge \mathrm{Re}(\mathfrak{f}(u,\zeta)\,d\zeta)$. We previously noted that $F \wedge F = 0$ is equivalent to $\mathbf{E} \cdot \mathbf{B} = 0$; see (5.32).

Exercise 5.8: Use the polar decomposition $\zeta = re^{i\theta}$ to show that the choice $\mathfrak{f}(u,\zeta) = f(u)/\zeta$, with $f(u)$ real, yields

$$F = \frac{f(u)}{r} du \wedge dr. \qquad (5.37)$$

The behaviour of the result in Exercise 5.8 is worth a discussion. The electric field in (5.37) always points along the radial direction, either directly away from or directly towards the $r = 0$ axis. Thus, (5.37) is invariant under a rotation of *any* angle around the $r = 0$ axis. This behaviour is very different to a linearly polarised plane wave in free space, which is only invariant under a rotation of 2π (or integer multiples of 2π). The electromagnetic 2-form (5.37) describes the fields in the vacuum between the inner and outer conductors of a coaxial waveguide.

5.5 OBSERVERS AND THE FIELDS THEY PERCEIVE

In the above discussion, the concept of the electromagnetic 2-form F emerged by drawing electric and magnetic fields into one mathematical object. The approach we took emphasised the relationship between the components of F and the components of the electric field and magnetic field associated with the basis dt, dx, dy, dz adapted to the Minkowski coordinate system t, x, y, z. However, from a geometrical perspective, this approach involves more mathematical structure than is strictly necessary. Although coordinates are often indispensable when calculating physical quantities, they are not a fundamental requirement for formulating physical theories. It is conceptually helpful to dispense with them and consider the meaning of "electric field" and "magnetic field" from a geometrical perspective.

To take the necessary steps, we need to introduce a timelike vector field V that corresponds to the four-velocities of a *continuum* of point-like observers. Evaluating V at a point (i.e. event) on the worldline of an observer produces a vector that is tangential to the worldline. A simple, familiar, example is $V = \partial/\partial t$, which is the field of four-velocities of inertial observers whose spatial coordinates x, y, z are constant. However, from a general perspective, the only requirements we have for V

are that it points towards the future (i.e. $dt(V) > 0$) and is timelike unit normalised (i.e. $g(V, V) = -1$).

The electric field $\mathcal{E}$ and magnetic field $\mathcal{B}$ perceived by observers whose four-velocities are given by V are

$$\mathcal{E} = \widetilde{i_V F}, \qquad \mathcal{B} = -\widetilde{i_V \star F} \qquad (5.38)$$

where the tilde denotes the map between vectors and 1-forms given by the spacetime metric g.

Exercise 5.9: Show

$$\mathcal{E} = E_x \frac{\partial}{\partial x} + E_y \frac{\partial}{\partial y} + E_z \frac{\partial}{\partial z}, \qquad \mathcal{B} = B_x \frac{\partial}{\partial x} + B_y \frac{\partial}{\partial y} + B_z \frac{\partial}{\partial z} \qquad (5.39)$$

follows from (5.15), (5.25), (5.31), (5.38) when $V = \partial/\partial t$.

Other than its aesthetic appeal, at first glance, it may seem that not much extra is gained from introducing (5.38) in addition to (5.15). However, there are computational advantages in decoupling the choice of coordinate system from the observer field V. A traditional method for setting up relativistic calculations often involves formulating the fields in a "locally inertial frame of reference" adapted to an observer, and transforming the results to the laboratory frame. However, this approach is somewhat cumbersome in comparison to a direct application of (5.38).

To appreciate (5.38), consider the electric field of a point particle at rest in the laboratory frame, as perceived by moving observers. In this case, the electromagnetic 2-form F is

$$F = \frac{q}{4\pi r^2} dt \wedge dr \qquad (5.40)$$

where $r = \sqrt{x^2 + y^2 + z^2}$ is the distance of the field point from the particle, and the particle has charge q. Furthermore, suppose that the observers are moving with (not necessarily constant) speed v along the z-axis. Thus

$$V = \frac{1}{\sqrt{1 - v^2}} \left(\frac{\partial}{\partial t} + v \frac{\partial}{\partial z} \right) \qquad (5.41)$$

and it follows

$$i_V F = \frac{q}{4\pi r^2} (dt(V) dr - dr(V) dt)$$

$$= \frac{q}{4\pi r^2} \frac{1}{\sqrt{1 - v^2}} \left(dr - \frac{vz}{r} dt \right) \qquad (5.42)$$

where, introducing $\gamma = 1/\sqrt{1 - v^2}$ for convenience, $dt(V) = Vt = \gamma$ and $dr(V) = Vr = \gamma v \, \partial r / \partial z = \gamma v z / r$ have been used in the final step.

The electric field is readily obtained by expressing the metric in terms of spherical polar coordinates. Introducing

$$x = r \sin \theta \cos \varphi, \qquad y = r \sin \theta \sin \varphi, \qquad z = r \cos \theta \qquad (5.43)$$

into (5.25) gives

$$g = -dt \otimes dt + dr \otimes dr + r^2 d\theta \otimes d\theta + r^2 \sin^2 \theta \, d\varphi \otimes d\varphi \qquad (5.44)$$

and so $\widetilde{\partial/\partial t} = -dt$ and $\widetilde{\partial/\partial r} = dr$. Since the metric dual of the metric dual is the identity map, we have the results $\widetilde{dt} = -\partial/\partial t$ and $\widetilde{dr} = \partial/\partial r$. Hence (5.38), (5.42) yield the electric field

$$\mathcal{E} = \frac{q}{4\pi r^2} \frac{1}{\sqrt{1 - v^2}} \left(v \cos \theta \frac{\partial}{\partial t} + \frac{\partial}{\partial r} \right) \qquad (5.45)$$

where $\cos \theta = z/r$ has been used. Note that (5.45) has a non-zero component along the t-axis, and this is a reflection of the fact that the four-velocities of the moving observers are not parallel to the four-velocities of observers at rest in the inertial frame of the particle.

Insight into the properties of (5.45) is gained by examining the magnitude $||\mathcal{E}||$ of the electric field. Since $||\mathcal{E}||^2 = g(\mathcal{E}, \mathcal{E})$ we have

$$||\mathcal{E}|| = \frac{|q|}{4\pi r^2} \sqrt{\frac{1 - v^2 \cos^2 \theta}{1 - v^2}}. \qquad (5.46)$$

Suppose, at an instant in time, the moving observer is at distance r from the particle. Inspection of (5.46) shows they will perceive a stronger electric field if they are in, for example, the xy-plane at $z = 0$ (so $\theta = \pi/2$) than if they are on the z-axis (i.e. $x = y = 0$, so $\theta = 0$).

5.6 ELECTRIC CHARGE AND ELECTRIC CURRENT FROM A FOUR-DIMENSIONAL PERSPECTIVE

The previous considerations focussed on the vacuum Maxwell equations. In a similar manner to the unification of electric and magnetic fields into one object (the electromagnetic 2-form F), the electric charge density ρ and electric current density $\mathbf{j}$ are different parts of the *electric current 3-form* j given by

$$\begin{aligned} j = -\rho \, dx \wedge dy \wedge dz \\ + j_x dt \wedge dy \wedge dz + j_y dt \wedge dz \wedge dx + j_z dt \wedge dx \wedge dy. \end{aligned} \qquad (5.47)$$

The Gauss-Ampère law is then

$$d \star F = j. \tag{5.48}$$

Note that the continuity equation

$$\frac{\partial \rho}{\partial t} + \boldsymbol{\nabla} \cdot \mathbf{j} = 0 \tag{5.49}$$

emerges from $dj = 0$, which itself follows immediately from (5.48) because the exterior derivative of the exterior derivative of any differential form is zero.

Exercise 5.10: Use (5.31) and (5.47) to verify that (5.48) is equivalent to

$$\boldsymbol{\nabla} \cdot \mathbf{E} = \rho, \qquad \boldsymbol{\nabla} \times \mathbf{B} = \frac{\partial \mathbf{E}}{\partial t} + \mathbf{j} \tag{5.50}$$

as expected.

5.6.1 Charge conservation

The differential identity $dj = 0$ is a concise, coordinate-free, way of expressing the fact that electric charge is conserved. The integral expression of charge conservation follows immediately by introducing a 4-dimensional region $\mathcal{U}_{(4)}$ in spacetime and making use of the generalised Stokes theorem

$$\int_{\mathcal{U}_{(4)}} dj = \int_{\partial \mathcal{U}_{(3)}} j \tag{5.51}$$

where $\partial \mathcal{U}_{(3)}$ is the 3-dimensional boundary of $\mathcal{U}_{(4)}$. The subscript has been introduced to emphasise the dimensionality of the integration domain.

Although $\mathcal{U}_{(4)}$ in (5.51) is completely general, it is helpful to consider a simple, specific, example to appreciate its physical meaning. Suppose $\mathcal{U}_{(4)}$ is the 4-dimensional region swept out by the interior of a ball of radius $r = r_0$ between the instants $t = t_i$ and $t = t_f$. Thus, the boundary $\partial \mathcal{U}_{(3)}$ of $\mathcal{U}_{(4)}$ consists of three "surfaces"[5] $\Sigma_{(3)}^i$, $\Sigma_{(3)}^f$, and $\Sigma_{(3)}^0$. The region

---

[5]We have used inverted commas to avoid confusion with the usual physical notion of a surface as a 2-dimensional region. However, from a mathematical perspective, a 3-dimensional region is a surface (technically, a *hypersurface*) in 4-dimensional spacetime.

$\Sigma^i_{(3)}$ is occupied by the ball at the instant $t = t_i$, the region $\Sigma^f_{(3)}$ is occupied by the ball at the instant $t = t_f$, and $\Sigma^0_{(3)}$ is the region swept out by the surface of the ball (i.e. the sphere $r = r_0$) between $t = t_i$ and $t = t_f$. Thus, (5.51) with $dj = 0$ leads to

$$0 = \int_{\Sigma^i_{(3)}} j + \int_{\Sigma^f_{(3)}} j + \int_{\Sigma^0_{(3)}} j. \tag{5.52}$$

Since $\Sigma^i_{(3)}$ and $\Sigma^f_{(3)}$ are both surfaces of constant t, the 1-form dt is zero when restricted to either of them. Hence, the only term in (5.47) that survives when j is restricted to $\Sigma^i_{(3)}$ or $\Sigma^f_{(3)}$ is $-\rho dx \wedge dy \wedge dz$. Likewise, the restriction of the 1-form dr to $\Sigma^0_{(3)}$ is zero because $\Sigma^0_{(3)}$ is a surface of constant r. Thus, to proceed further, it is useful to express (5.47) in terms of spherical polar coordinates (5.43). Note that, since dr is zero when restricted to $\Sigma^0_{(3)}$, it is permissible to set $r = r_0$ in (5.43) before applying the exterior derivative inside integrals over $\Sigma^0_{(3)}$.

Exercise 5.11: Substitute $r = r_0$ in (5.43) and show

$$dx \wedge dy = r_0^2 \sin \theta \cos \theta \, d\theta \wedge d\varphi, \tag{5.53}$$
$$dz \wedge dx = r_0^2 \sin^2 \theta \sin \varphi \, d\theta \wedge d\varphi, \tag{5.54}$$
$$dy \wedge dz = r_0^2 \sin^2 \theta \cos \varphi \, d\theta \wedge d\varphi. \tag{5.55}$$

Hence, argue (5.47) can be expressed as

$$j = (j_x \sin \theta \cos \varphi + j_y \sin \theta \sin \varphi + j_z \cos \theta) r^2 \sin \theta \, dt \wedge d\theta \wedge d\varphi + \dots \tag{5.56}$$

where $\dots$ indicates terms that must contain dr.

Thus, (5.52) can be written as

$$0 = -\int_{\Sigma^i_{(3)}} \rho r^2 \sin \theta \, dr \wedge d\theta \wedge d\varphi - \int_{\Sigma^f_{(3)}} \rho r^2 \sin \theta \, dr \wedge d\theta \wedge d\varphi$$
$$+ \int_{\Sigma^0_{(3)}} j_r r_0^2 \sin \theta \, dt \wedge d\theta \wedge d\varphi \tag{5.57}$$

where

$$j_r = j_x \sin \theta \cos \varphi + j_y \sin \theta \sin \varphi + j_z \cos \theta \tag{5.58}$$

and

$$dx \wedge dy \wedge dz = r^2 \sin \theta \, dr \wedge d\theta \wedge d\varphi \tag{5.59}$$

have been used.

The final step in reducing (5.57) to ordinary integrals is to account for the relative signs that arise due to orientation. The 3-dimensional integration domains $\Sigma^i_{(3)}$, $\Sigma^f_{(3)}$, $\Sigma^0_{(3)}$ are ingredients in the boundary of the 4-dimensional region $\mathcal{U}_{(4)}$, and the first step in determining a consistent choice of relative signs is to select a 4-form. The spacetime volume 4-form $\star 1$ is a natural choice.

There are two possible sources of additional minus signs when erasing the wedges. The first consideration is the position of the 1-form in $\star 1$ that *is not* inside a given integral. If the 1-form is in an odd position then the sign is positive, whereas it is negative if the 1-form is in an even position. In particular, since

$$\star 1 = dt \wedge dx \wedge dy \wedge dz \tag{5.60}$$
$$= r^2 \sin \theta \, dt \wedge dr \wedge d\theta \wedge d\varphi \tag{5.61}$$

we see that $dr \wedge d\theta \wedge d\varphi \mapsto +dr \, d\theta \, d\varphi$ because dt is missing, whereas $dt \wedge d\theta \wedge d\varphi \mapsto -dt \, d\theta \, d\varphi$ because dr is missing. The second consideration is whether the integration domain is at the start or end of an interval. An overall minus sign is introduced if the domain is at the start of the interval. Thus, because t increases from t_i to t_f, we must make the replacements $\int_{\Sigma^i_{(3)}} \mapsto -\int_{r=0}^{r=r_0} \int_{\theta=0}^{\theta=\pi} \int_{\varphi=0}^{\varphi=2\pi}$ and $\int_{\Sigma^f_{(3)}} \mapsto +\int_{r=0}^{r=r_0} \int_{\theta=0}^{\theta=\pi} \int_{\varphi=0}^{\varphi=2\pi}$. Likewise, since r increases from 0 to r_0 we have the replacement $\int_{\Sigma^0_{(3)}} \mapsto +\int_{t=t_i}^{t=t_f} \int_{\theta=0}^{\theta=\pi} \int_{\varphi=0}^{\varphi=2\pi}$. Thus, the explicit expression of (5.57) in terms of ordinary nested integrals is

$$\int_0^{2\pi} \int_0^{\pi} \int_0^{r_0} \rho \, r^2 \sin \theta \, dr \, d\theta \, d\varphi \bigg|_{t=t_i} - \int_0^{2\pi} \int_0^{\pi} \int_0^{r_0} \rho \, r^2 \sin \theta \, dr \, d\theta \, d\varphi \bigg|_{t=t_f}$$
$$= \int_0^{2\pi} \int_0^{\pi} \int_{t_i}^{t_f} j_r \, r^2 \sin \theta \, dt \, d\theta \, d\varphi \bigg|_{r=r_0}. \tag{5.62}$$

Exercise 5.12: Derive (5.62) from (5.49) using standard techniques in vector calculus, without making use of differential forms.

The terms on the left-hand side of (5.62) combine to give the drop in the total charge inside the ball between the initial and final times. The right-hand side of (5.62) is the total current flowing outwards through the surface of the ball.

Note that if we had considered a hollow ball instead of a solid ball then there would also be an integral over the hollow ball's inner surface in (5.52). The approach used to determine the overall relative sign of its contribution to (5.62) would be the same as we used to fix the relative signs of the integrals at constant t. When expressed in terms of ordinary nested integrals rather than an integral over a differential form, the extra term that emerges would have the opposite sign to the term on the right-hand side of (5.62) because r increases from the inner radius to the outer radius in the integrals over the charge density.

From a physical perspective, it is obvious that if the electric current is zero at sufficiently large r_0, so no charge flows into or out from the ball, then the total charge inside the ball remains constant. It is illuminating to see how this conclusion emerges from very general, conceptual, considerations based on (5.51) and $dj = 0$. Suppose that $\Sigma^{I}_{(3)}$, $\Sigma^{I'}_{(3)}$ are two 3-dimensional regions that can be smoothly deformed into each other, i.e. without making changes that require, for example, tearing and gluing.

Now consider a 4-dimensional region $\mathcal{U}_{(4)}$ whose boundary $\partial \mathcal{U}_{(3)}$ contains the regions $\Sigma^{I}_{(3)}$, $\Sigma^{II}_{(3)}$ and a third (in general, multi-component) region $\Sigma^{III}_{(3)}$. See Figure 5.1. Now, if j is zero on $\Sigma^{III}_{(3)}$ then (5.51) and $dj = 0$ give

$$0 = \int_{\Sigma^{I}_{(3)}} j + \int_{\Sigma^{II}_{(3)}} j. \tag{5.63}$$

Likewise, consider a 4-dimensional region $\mathcal{U}'_{(4)}$ whose boundary $\partial \mathcal{U}'_{(3)}$ contains the regions $\Sigma^{I'}_{(3)}$, $\Sigma^{II'}_{(3)}$, and $\Sigma^{III'}_{(3)}$. If j is zero on $\Sigma^{III'}_{(3)}$ then we arrive at

$$0 = \int_{\Sigma^{I'}_{(3)}} j + \int_{\Sigma^{II'}_{(3)}} j. \tag{5.64}$$

Thus, if we choose $\mathcal{U}'_{(4)}$ such that $\Sigma^{II'}_{(3)} = \Sigma^{II}_{(3)}$ then comparison of (5.63) and (5.64) leads to $\int_{\Sigma^{I}_{(3)}} j = \int_{\Sigma^{I'}_{(3)}} j$. Another way of stating this result is as follows. Let $\Sigma^{\lambda}_{(3)}$ be a family of 3-dimensional regions (i.e. each choice of λ corresponds to a 3-dimensional region) whose dependence on the parameter λ is smooth. It follows that the quantity given by

$$Q[\Sigma^{\lambda}_{(3)}] = \int_{\Sigma^{\lambda}_{(3)}} j \tag{5.65}$$

has the same value for all λ in a given interval, as long as j is zero outside of a region fully contained in $\Sigma^{\lambda}_{(3)}$. In particular, if λ is identified

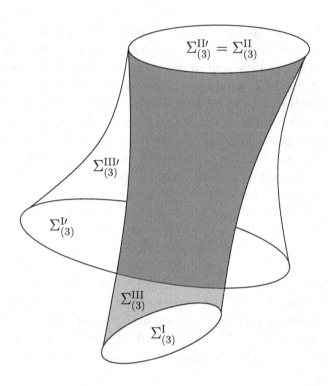

Figure 5.1 A schematic diagram of two 4-dimensional regions whose boundaries share a common component. The components of the boundaries of $\mathcal{U}_{(4)}$ and $\mathcal{U}'_{(4)}$ are labelled $\Sigma^{\mathrm{I}}_{(3)}$, $\Sigma^{\mathrm{II}}_{(3)}$, $\Sigma^{\mathrm{III}}_{(3)}$ and $\Sigma^{\mathrm{I}\prime}_{(3)}$, $\Sigma^{\mathrm{II}\prime}_{(3)}$, $\Sigma^{\mathrm{III}\prime}_{(3)}$, respectively. The region $\Sigma^{\mathrm{III}}_{(3)}$ has been shaded for clarity, and a darker shade has been used for the portion of $\Sigma^{\mathrm{III}}_{(3)}$ inside $\mathcal{U}'_{(4)}$.

with time t then the constancy of (5.65) is readily interpreted as charge conservation. Since j is zero outside of a region fully contained in $\Sigma^{\lambda}_{(3)}$, no charge flows into or out of the region. However, note that, in general, there is no requirement for $\Sigma^{\lambda}_{(3)}$ to be a surface of constant t. For example, it could be a surface of constant $u = x - t$, i.e. a null surface, or have different properties at different events in spacetime. The above construction is very general, and it illustrates some of the benefits of thinking in a geometrical (i.e. coordinate-free) manner.

5.7 POLARISATION AND MAGNETISATION

The Gauss-Ampère law (5.48) and the Gauss-Faraday law (5.22) are commonly regarded as a *microscopic* (albeit classical) formulation of

Maxwell's equations, in which the electric charges and currents are represented entirely by the 3-form j. However, for many physical applications, it is helpful to think in terms of *macroscopic* quantities that arise by locally averaging (i.e. coarse-graining) the fields over fast time scales and short distances associated with the microscopic behaviour. This provides a practical way of representing the macroscopic electromagnetic properties of a vast number of interacting particles in a material whose internal behaviour features widely differing scales.

The locally averaged fields satisfy the macroscopic version of Maxwell's equations

$$\nabla \cdot \langle \mathbf{B} \rangle = 0, \qquad \nabla \times \langle \mathbf{E} \rangle = -\frac{\partial \langle \mathbf{B} \rangle}{\partial t}, \tag{5.66}$$

$$\nabla \cdot \langle \mathbf{E} \rangle = \langle \rho \rangle, \qquad \nabla \times \langle \mathbf{B} \rangle = \frac{\partial \langle \mathbf{E} \rangle}{\partial t} + \langle \mathbf{j} \rangle \tag{5.67}$$

where the angle brackets $\langle \ldots \rangle$ denote a locally averaged quantity. The averaged charge density $\langle \rho \rangle$ and averaged current density $\langle \mathbf{j} \rangle$ are separated into *free* and *bound* parts,

$$\langle \rho \rangle = \rho_{\text{free}} + \rho_{\text{bound}}, \qquad \langle \mathbf{j} \rangle = \mathbf{j}_{\text{free}} + \mathbf{j}_{\text{bound}}, \tag{5.68}$$

where the "free" charge density and current density describe the "bulk" behaviour of the relatively mobile charges in the material (e.g. the electrons that flow from atom to atom), whereas the "bound" contributions encode the localised behaviour of the charge and current distributions within the individual atoms in the material. The "bound" contributions are written in terms of the electric polarisation $\mathbf{P}$ and magnetisation $\mathbf{M}$ induced as responses to the averaged fields $\langle \mathbf{E} \rangle$ and $\langle \mathbf{B} \rangle$. In particular,

$$\rho_{\text{bound}} = -\nabla \cdot \mathbf{P}, \qquad \mathbf{j}_{\text{bound}} = \frac{\partial \mathbf{P}}{\partial t} + \nabla \times \mathbf{M} \tag{5.69}$$

which introduced into (5.67) gives

$$\nabla \cdot \mathbf{D} = \rho_{\text{free}}, \qquad \nabla \times \mathbf{H} = \frac{\partial \mathbf{D}}{\partial t} + \mathbf{j}_{\text{free}} \tag{5.70}$$

where

$$\langle \mathbf{E} \rangle = \mathbf{D} - \mathbf{P}, \qquad \langle \mathbf{B} \rangle = \mathbf{H} + \mathbf{M}. \tag{5.71}$$

The spacetime version of the above considerations follows the same reasoning. The locally averaged electromagnetic 2-form $\langle F \rangle$ and electric current 3-form $\langle j \rangle$ satisfy the macroscopic versions

$$d\langle F \rangle = 0, \qquad d \star \langle F \rangle = \langle j \rangle \tag{5.72}$$

of the microscopic Gauss-Faraday law (5.22) and microscopic Gauss-Ampère law (5.48), where

$$\langle F \rangle = \langle E_x \rangle \, dt \wedge dx + \langle E_y \rangle \, dt \wedge dy + \langle E_z \rangle \, dt \wedge dz$$
$$- \langle B_x \rangle \, dy \wedge dz - \langle B_y \rangle \, dz \wedge dx - \langle B_z \rangle \, dx \wedge dy \quad (5.73)$$

and

$$\langle j \rangle = - \langle \rho \rangle \, dx \wedge dy \wedge dz$$
$$+ \langle j_x \rangle dt \wedge dy \wedge dz + \langle j_y \rangle dt \wedge dz \wedge dx + \langle j_z \rangle dt \wedge dx \wedge dy. \quad (5.74)$$

The 3-form $\langle j \rangle$ is expressed as

$$\langle j \rangle = j_{\text{free}} - d \star \Pi \quad (5.75)$$

where the 3-form j_{free} represents the "free" charge and current, and the 2-form Π is the spacetime amalgam

$$\Pi = P_x \, dt \wedge dx + P_y \, dt \wedge dy + P_z \, dt \wedge dz$$
$$+ M_x \, dy \wedge dz + M_y \, dz \wedge dx + M_z \, dx \wedge dy \quad (5.76)$$

of the polarisation **P** and magnetisation **M**. Thus, the macroscopic Gauss-Ampère law can be written as

$$d \star G = j_{\text{free}} \quad (5.77)$$

where the 2-form

$$G = \langle F \rangle + \Pi \quad (5.78)$$

incorporates the effects of $\langle F \rangle$ on the "bound" charge and current. The expression for G in terms of **D** and **H** is

$$G = D_x \, dt \wedge dx + D_y \, dt \wedge dy + D_z \, dt \wedge dz$$
$$- H_x \, dy \wedge dz - H_y \, dz \wedge dx - H_z \, dx \wedge dy. \quad (5.79)$$

Exercise 5.13: A linear isotropic material satisfies the electromagnetic constitutive equations $\mathbf{D} = \epsilon_r \langle \mathbf{E} \rangle$ and $\mathbf{H} = \langle \mathbf{B} \rangle / \mu_r$, where the scalar field ϵ_r is the relative permittivity of the material and the scalar field μ_r is its relative permeability. Use (5.79) and (5.73) to show

$$G = \frac{\langle F \rangle}{\mu_r} + \left(\epsilon_r - \frac{1}{\mu_r} \right) dt \wedge i_{\partial_t} \langle F \rangle. \quad (5.80)$$

We previously explained how one can determine the electric and magnetic fields perceived by a continuum of observers. We represent the observers using the future-directed, timelike unit normalised vector field V. The result of evaluating the field V at an event in spacetime is the four-velocity of the observer at that event. Thus, (5.80) can be readily generalised to accommodate moving materials. We simply identify the field V with the (average) four-velocities of the atoms of the material.

The simplest approach is to assume that the electromagnetic constitutive equations for a linear isotropic material hold in the instantaneous rest frame of the atoms of the material. Thus,

$$G = \frac{\langle F \rangle}{\mu_r} - \left(\epsilon_r - \frac{1}{\mu_r} \right) \tilde{V} \wedge i_V \langle F \rangle \qquad (5.81)$$

where the extra sign arises because $\tilde{V} = -dt$ if $V = \partial/\partial t$; see (5.25). In principle, one could include extra terms in (5.81) that depend on the four-acceleration of the atoms. In asserting (5.81), we are neglecting the effects of non-inertial forces on the relative orientations of the atoms and the shapes of their electronic orbitals. Even so, derivatives of V emerge when G is plugged into (5.77), and so the electric and magnetic fields in the instantaneous rest frame of an accelerating material behave differently from those that are supported by the material when it is at rest.

CHAPTER 6

Classical mechanics

One of the most important areas of physics is classical mechanics. In this chapter, we explore how differential geometry can facilitate formulations of classical mechanics that are both elegant and powerful. This will lead us to the rich structures of *tangent bundles* and *cotangent bundles*. This is just a glimpse into the theory of fibre bundles, which underlie a great many fields within physics and beyond.

However, before commencing our journey, we need to introduce a little more terminology. Thus far, to avoid straying too far from familiar territory, we have avoided using the word "manifold". This approach is reasonable because the spaces we have considered so far are very familiar in undergraduate physics. We only considered the space of thermodynamic states or the space (or spacetime) in which physical objects exist, and focussed on the operational aspects of exterior calculus and tensor calculus. However, we are now going to consider more general settings and, in such circumstances, it is conventional to be a little more precise.

The detailed definition of *manifold* is beyond the scope of this book, but it is sufficient for our purposes to regard it as an abstract set of points that can be regarded as "smooth", in the sense that it can be completely covered by one or more coordinate systems. A *submanifold* is a subset of a manifold that is itself a manifold. All of the spaces we have considered so far (Euclidean space, Minkowski spacetime and thermodynamic state space) are examples of manifolds. The surface of a ball in 3-dimensional Euclidean space is an example of a submanifold.

6.1 THE TANGENT BUNDLE

Consider an n-dimensional manifold M. As in previous chapters, M may represent physical space or a more abstract configuration space. We have

DOI: 10.1201/9781003228943-6

seen that vectors at a point $p \in M$ lie not in M itself, but in the tangent space $T_p M$ at $p \in M$. $T_p M$ is an n-dimensional vector space. To identify a particular vector v, then, we must specify $2n$ pieces of information: n coordinates to identify the point $p \in M$, and a further n components of the vector $v \in T_p M$. The set of all tangent vectors to a space M is itself a new space, with twice the dimension of the original. We call this space the *tangent bundle* to M, and write it

$$TM = \bigcup_{p \in M} T_p M. \tag{6.1}$$

This tangent bundle is a $2n$-dimensional manifold, consisting of every possible vector at every point of the original manifold M.

The tangent bundle is a manifold in its own right, and as such we are free to use any coordinates on it. However, a coordinate system $\{x^a\}$ on (some region of) M induces a natural coordinate system on (a corresponding region of) TM. Since $\{\partial_a\}$ forms a basis for $T_p M$, an arbitrary vector v at $p \in M$ can be written $v = v^a \partial_a$ (employing the Einstein summation convention, and the abbreviation $\partial_a = \partial/\partial x^a$). Then the set $\{x^a, v^b\}$ determines a unique combination of a point on M and a vector at that point, and hence can be used as coordinates for TM.

The construction of TM equips it with a map π that projects from TM to M:

$$\begin{aligned} \pi: \quad & TM \to M, \\ & (p, v) \mapsto p. \end{aligned} \tag{6.2}$$

This corresponds to keeping the point in M, but "forgetting" about the choice of vector. It should be noted that there is no equivalent map that preserves the vector, but forgets about the point in M. This is because we cannot in general equate two vectors belonging to different tangent spaces.[1]

What we have called a *vector field* V can be seen as a submanifold of TM, such that the action of π restricted to this submanifold is both injective (no two elements of V are mapped to the same $p \in M$) and

---

[1]This may seem surprising, since in everyday life we often relate, say, velocities at different positions. This is possible because flat Euclidean space is unusual in allowing vectors to be transported from one position to another in a way that does not depend on the path taken.

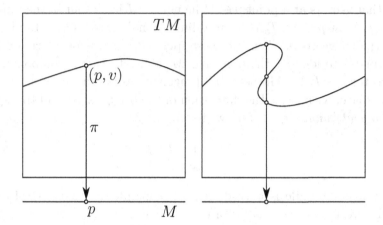

Figure 6.1 Two schematic diagrams of the tangent bundle TM and its properties. The labels in the diagram on the right have been suppressed for clarity. The diagram on the left shows a section of TM. Note that the projection map π associates each point p in M with only one point (p, v) in the section. The diagram on the right shows a submanifold of TM that is not a section, because not every point in M corresponds to only one point in the submanifold.

surjective (every $p \in M$ is the projection of some point on the submanifold). Such a submanifold is known as a *section* of TM. Figure 6.1 shows an example of a section and a submanifold that is not a section.

It is also possible to map certain structures from M to TM; such maps are known as *lifts*. For example, a function $f(p)$ on M can be lifted to a function $F(p, v)$ on TM, satisfying $F(p, v) = f(p)$ (this holds regardless of the value of v, i.e. $F(p, w) = F(p, v)$). With abuse of notation, and providing no confusion can arise, we will sometimes use the same symbol for both functions, e.g. $f(p, v) = f(p)$.

In a similar way, it is possible to lift differential forms from M to TM. A general 1-form on TM can be written $\Omega = \omega_a dx^a + \Omega_a dv^a$. Then the lift of a 1-form $\alpha = \alpha_a dx^a$ on M is given by $\omega_a = \alpha_a$, $\Omega_a = 0$, where the functions $\{\alpha_a\}$ are lifted to TM as described above. The generalisation to arbitrary p-forms is straightforward.

There is an alternative way to lift a 1-form to the tangent bundle, where this time it results in a scalar (not a 1-form) on TM. Consider again the 1-form α on M. By lifting its components to TM, we can construct the 0-form $\alpha^* = \alpha_a v^a$. We refer to α^* as the *scalar lift of* α.[2]

The scalar lift generalises to other rank $(0, m)$ tensors in the obvious way: coefficients are lifted as scalar functions, whilst basis covectors are replaced by vector components: $dx^a \mapsto v^a$. Thus the lift of a scalar function $F(p, v) = f(p)$ can be interpreted as the scalar lift f^* (where we have suppressed the argument (p) for notational simplicity).

Exercise 6.1: For a 2-form β, show that $\beta^* = 0$. Determine the scalar lift g^* of the metric tensor $g = g_{ab}\, dx^a \otimes dx^b$.

Finally, and most importantly for our purposes, consider a curve on M,

$$\gamma : \mathbb{R} \to M,$$
$$\lambda \mapsto \gamma(\lambda). \tag{6.3}$$

This represents a path on the configuration space. We define the lift of this curve to TM as

$$\Gamma : \mathbb{R} \to TM,$$
$$\lambda \mapsto \left(\gamma(\lambda), \gamma'(\lambda)\right), \tag{6.4}$$

where $\gamma'(\lambda)$ is the tangent to $\gamma(\lambda)$. $\Gamma = (\gamma, \gamma')$ is now a path not on the configuration space, but on the tangent bundle. It is important to note that this lifted curve depends not only on the *image* of γ, but on its parameterisation: reparameterising γ will not merely reparameterise Γ, but lead to a different curve on TM.

6.2 LAGRANGIAN MECHANICS

Having introduced the essential features of the tangent bundle, we can now explore its role in an elegant formulation of classical mechanics, known as *Lagrangian mechanics*.

At any instant, the state of a physical system (e.g. positions of a set of particles) is described by a position in configuration space, M.

---

[2]Neither the name "scalar lift" nor the notation α^* is standard terminology for this construction, which appears not to have a widely accepted nomenclature in the literature.

As the system evolves, its state traces out a curve γ on M. The evolution parameter will usually be time (for relativistic systems there is a choice between coordinate time and proper time), but not necessarily. In accelerator science, for example, it is often convenient to describe evolution with respect to distance z along some idealised *design orbit*. In the following, we denote the evolution parameter t, without prejudicing its interpretation.

Given a parameterised curve $\gamma(t)$ on M, we lift it to a curve $\Gamma(t)$ on TM. The goal of Lagrangian mechanics is to determine $\Gamma(t)$.

The *Lagrangian* is a function $L(p, v)$ on TM. It is important to recognise that this is not the lift of some function on M, but is intrinsically defined on TM itself. Given this Lagrangian, we can integrate it along Γ, to yield the *action*,

$$S[\Gamma] = \int_\Gamma L dt. \tag{6.5}$$

We explicitly include Γ as an argument of S, to emphasise that the action is a *functional* of the curve: it takes a curve as input, and outputs a number. If we alter the curve $\Gamma \to \hat{\Gamma}$, the action will change $S[\Gamma] \to S[\hat{\Gamma}]$.

The fundamental principle of Lagrangian mechanics is *Hamilton's principle*, which states that the path taken by the physical system is that which makes the action stationary. In other words, small deformations of Γ leave the action unchanged, i.e. $\delta S := S[\hat{\Gamma}] - S[\Gamma] = 0$.

To use Hamilton's principle, we let $\{x^a, v^b\}$ denote the coordinates of Γ, and $\{x^a + \delta x^a, v^b + \delta v^b\}$ those of $\hat{\Gamma}$. Then we have

$$\delta S = \int_\Gamma \left(\frac{\partial L}{\partial x^a} \delta x^a + \frac{\partial L}{\partial v^a} \delta v^a \right) dt. \tag{6.6}$$

If we require that not only Γ but also $\hat{\Gamma}$ are lifts of curves from M, then $\delta v^a = d\delta x^a / dt$, and integrating the second term in (6.6) by parts yields

$$\delta S = \frac{\partial L}{\partial v^a} \delta x^a \Big|_{\partial \Gamma} + \int_\Gamma \left(\frac{\partial L}{\partial x^a} - \frac{d}{dt} \frac{\partial L}{\partial v^a} \right) \delta x^a dt, \tag{6.7}$$

where $\partial \Gamma$ is the boundary of Γ, i.e. the start and end points. Restricting to curves $\hat{\Gamma}$ that coincide with Γ at the start and end points, $\delta x^a |_{\partial \Gamma} = 0$, and the first term vanishes. Since δx^a is otherwise arbitrary, it follows that the curve Γ must satisfy the *Euler-Lagrange equations*,

$$\frac{\partial L}{\partial x^a} - \frac{d}{dt} \frac{\partial L}{\partial v^a} = 0. \tag{6.8}$$

Note that there are n such equations, where n is the dimension of M.

Before considering some examples, we remark on a more general property of the Euler-Lagrange equations. If L has no dependence on a given coordinate, say x^1, then (6.8) reduces to

$$\frac{d}{dt}\frac{\partial L}{\partial v^1} = 0 \quad \Longrightarrow \quad \frac{\partial L}{\partial v^1} = \text{const.} \tag{6.9}$$

Thus any invariance of the Lagrangian corresponds to a *conserved quantity*. We will explore this phenomenon in more detail in Section 6.6, in the context of Hamiltonian mechanics.

6.2.1 Example: Free particle

The simplest physical system is perhaps the non-relativistic free particle. For this case, the Lagrangian is simply the *kinetic energy*. From basic physics, we know this should be "half mass times velocity squared". In terms of intrinsic geometric quantities, we can express this as

$$L = \frac{1}{2}mg^* = \frac{1}{2}mg_{ab}v^a v^b. \tag{6.10}$$

In 2-dimensional flat Euclidean space in Cartesian coordinates, then, this becomes

$$L = \frac{1}{2}m(v_x^2 + v_y^2), \tag{6.11}$$

and the Euler-Lagrange equations yield

$$\dot{v}_x = 0, \qquad \dot{v}_y = 0, \tag{6.12}$$

where a dot represents differentiation with respect to the evolution parameter t. In keeping with our general observation, this Lagrangian has no dependence on two coordinates, and we have two constants of the motion, v_x and v_y.

It is worth considering the same example again, but this time in terms of polar coordinates r and θ. This time, (6.10) becomes

$$L = \frac{1}{2}m(v_r^2 + r^2 v_\theta^2), \tag{6.13}$$

leading to the Euler-Lagrange equations in the form

$$\dot{v}_r = rv_\theta^2, \qquad r^2 v_\theta = a, \tag{6.14}$$

with a a constant of the motion.

In this case, the Lagrangian is independent of just one coordinate, θ, yielding just one conserved quantity, a. However, since this is an alternative description of the same Lagrangian (6.10), they must lead to the same dynamics, including the same number of conserved quantities. A second conserved quantity can be seen to be the *speed*,

$$V = \sqrt{g^*} = \sqrt{v_r^2 + r^2 v_\theta^2}. \qquad (6.15)$$

If such a "hidden" constant of the motion can be found, it can replace one of the Euler-Lagrange equations (in this case, the equation for v_r).

Exercise 6.2: Show that V does not vary in time. Use the constants a and V to express the velocities v_r and v_θ as functions of r. Is there any ambiguity in the resulting expressions?

6.2.2 Example: Harmonic oscillator

Looking beyond the free particle, there are many cases where the Lagrangian can be written

$$L = \frac{1}{2} m g^* - \mathcal{V}, \qquad (6.16)$$

where the *potential energy* $\mathcal{V}$ is the scalar lift of a 0-form on the configuration space M. Then the Euler-Lagrange equations become

$$m \frac{d(g_{ab} v^b)}{dt} = -\frac{\partial \mathcal{V}}{\partial x^a}. \qquad (6.17)$$

Unsurprisingly, this is reminiscent of Newton's second law of motion $\mathbf{F} = m\mathbf{a}$, with a conservative force $\mathbf{F} = -\nabla \mathcal{V}$, in classical vector calculus notation. However, the two are only directly equivalent for flat space in Cartesian coordinates. In general, the left-hand side of (6.17) need not correspond to components of kinetic momentum, nor the right-hand side to components of a force.

To illustrate the use of (6.16), we turn to the prototypical example of the 1-dimensional harmonic oscillator. In this case, $g^* = v^2$, and $\mathcal{V} = kx^2/2$:

$$L = \frac{1}{2} m v^2 - \frac{1}{2} k x^2. \qquad (6.18)$$

Being 1-dimensional, there is just a single Euler-Lagrange equation,

$$m\dot{v} = -kx. \tag{6.19}$$

Recognising that the velocity is the derivative of the coordinate (since $\Gamma(t)$ is the lift of a curve on M), this can be recast as the familiar equation of simple harmonic motion,

$$\ddot{x} + \omega^2 x = 0, \tag{6.20}$$

with its well-known solutions $x = A\cos(\omega t + \phi)$, where the angular frequency is defined by $\omega = \sqrt{k/m}$, and A and ϕ are constants.

For a general $\mathcal{V}$, the dynamics is not so readily solved. However, despite the Lagrangian's dependence on the coordinate x, we can nevertheless find a conserved quantity. This time, it is the *total energy* $\frac{1}{2}mg^* + \mathcal{V}$ that is conserved.

Exercise 6.3: Demonstrate that the total energy is conserved.

6.2.3 Example: Particle in a magnetic field

Not all dynamical systems are described by Lagrangians of the form $L = \frac{1}{2}mg^* - \mathcal{V}$. As a final example, we consider a point particle of charge q responding to a magnetic field. This is described in terms of the magnetic 1-form potential A. The Lagrangian is

$$L = \frac{1}{2}mg^* + qA^* = \frac{1}{2}mg_{ab}v^a v^b + qA_a v^a. \tag{6.21}$$

Assuming $\{x^a\}$ are Cartesian coordinates, the Euler-Lagrange equations then read

$$\frac{d(mg_{ab}v^b + qA_a)}{dt} = q\frac{\partial A_b}{\partial x^a}v^b. \tag{6.22}$$

To make sense of this, we need to understand how to differentiate the functions A_a with respect to t. From the definition of t, d/dt is the derivative along Γ. Hence the derivative of a function f along Γ on TM is given by

$$\frac{df}{dt} = \frac{\partial f}{\partial x^a}\dot{x}^a + \frac{\partial f}{\partial v^a}\dot{v}^a. \tag{6.23}$$

Since $\Gamma(t)$ is the lift of a curve $\gamma(t)$ on M, $\dot{x}^a = v^a$, while $\dot{v}^a$ can (usually) be determined from the Euler-Lagrange equations. However, since the functions A_a are lifts of functions on M, $\partial A_a / \partial v^b = 0$, and we have

$$\frac{dA_a}{dt} = \frac{\partial A_a}{\partial x^b} v^b. \tag{6.24}$$

Using this in (6.22) gives

$$m \frac{d(g_{ab} v^b)}{dt} = q v^b \left(\frac{\partial A_b}{\partial x^a} - \frac{\partial A_a}{\partial x^b} \right). \tag{6.25}$$

The terms in brackets are the components of dA, which we recognise as the magnetic field 2-form B (i.e. $B = -F$ in the absence of electric fields; see (5.15)), lifted to TM. Hence we recognise the right-hand side of (6.25) as the components of the Lorentz force on a particle of charge q in a magnetic field $B = dA$.

6.3 THE COTANGENT BUNDLE

In the previous sections, we saw that the tangent spaces at all points of a manifold M can be united into the *tangent bundle TM*, leading to an elegant formulation of classical mechanics. A natural question then is whether a similar structure can be developed from the cotangent spaces, and if so whether it can have applications in physics. The answer to both questions, as we will see in the remainder of this chapter, is emphatically yes.

As we did for the tangent spaces, we can assemble all the cotangent spaces into the *cotangent bundle*

$$T^\star M = \bigcup_{r \in M} T_r^\star M. \tag{6.26}$$

This cotangent bundle is again a $2n$-dimensional manifold, this time consisting of every possible covector at every point of the original manifold M.

We saw that a coordinate system on M induces a natural coordinate system on TM, and in a very similar way it can induce a coordinate system on $T^\star M$. When relating the tangent and cotangent bundles (as we will in the next section) it is convenient to employ different labels for coordinates on M in these different contexts. Hence we here use coordinates $\{q^a\}$ for M, to distinguish from the $\{x^a\}$ coordinates used in the discussions of the tangent bundle.

Given coordinates $\{q^a\}$ on M, $\{dq^a\}$ forms a basis for $T_r^\star M$. An arbitrary covector at $r \in M$ can therefore be written $p_a dq^a$. Then the set $\{q^a, p_b\}$ determines a unique combination of a point on M and a covector at that point, and hence can be used as coordinates for $T^\star M$.

As with TM, there is a natural map π that projects from $T^\star M$ to M:

$$\pi : \quad T^\star M \to M,$$
$$(q^a, p_b) \mapsto (q^a). \tag{6.27}$$

We could use different symbols to distinguish the projection maps from TM and from $T^\star M$, but this is not necessary: the usage should be clear from context. Once again, π corresponds to keeping the point in M, but "forgetting" about the choice of covector.

So far, the reader could be forgiven for thinking our treatment of $T^\star M$ is merely replicating the structures we encountered on TM. However, the cotangent bundle has a richer structure than the tangent bundle. Its formulation immediately endows it with a 1-form, known variously as the *Poincaré, canonical,* or *tautological* 1-*form,*

$$\theta = p_a dq^a. \tag{6.28}$$

It is important not to mistake θ for a 1-form on M: the p's here are coordinates on $T^\star M$, not constants or functions of the q's. Thus the Poincaré 1-form is a differential form on $T^\star M$.

Exercise 6.4: The Poincaré 1-form is expressed in (6.28) in terms of coordinates. Show that it is well defined, i.e. under a change of coordinates on M, θ retains its form. Is the same true of the analogous form[3] $\sum_{a=1}^{n} v^a dx^a$ on the tangent bundle?

The Poincaré 1-form can be used to naturally endow $T^\star M$ with a *symplectic structure.* A symplectic manifold is any manifold that possesses a closed, non-degenerate 2-form, which we call the *symplectic 2-form.* Such a 2-form is given by (the negative of) the exterior derivative

---

[3]We explicitly indicate the sum in this expression, as the Einstein summation convention applies only when the repeated index appears once "up" and once "down". The reader is advised to be wary of expressions with indices repeated twice up, or twice down.

of the Poincaré 1-form:

$$\omega = -d\theta = dq^a \wedge dp_a. \tag{6.29}$$

That ω is closed follows immediately from its definition as the exterior derivative of a 1-form, and the vanishing of repeated application of the exterior derivative.

For a 2-form to be non-degenerate means that its interior contraction with a vector X vanishes only if the vector is zero:

$$i_X\omega = 0 \iff X = 0. \tag{6.30}$$

An arbitrary vector X on $T^\star M$ can be written

$$X = \xi^a \frac{\partial}{\partial q^a} + \zeta_a \frac{\partial}{\partial p_a}, \tag{6.31}$$

and its contraction with ω as

$$i_X\omega = \xi^a dp_a - \zeta_a dq^a, \tag{6.32}$$

where we have used $dq^a(\partial/\partial q^b) = \delta^a_b = dp_b(\partial/\partial p_a)$, and $dq^a(\partial/\partial p_b) = 0 = dp_b(\partial/\partial q^a)$. Since the dq's and dp's are independent, $i_X\omega = 0$ if and only if $\xi^a = \zeta_b = 0$, i.e. $X = 0$. Hence ω is non-degenerate.

We can use ω to associate a 1-form,

$$X^\flat = i_X\omega, \tag{6.33}$$

to each vector field X, and a vector field $\alpha^\sharp$,

$$\alpha = i_{\alpha^\sharp}\omega, \tag{6.34}$$

to each 1-form α, in much the same way as we did using the metric tensor.

Exercise 6.5: Equation (6.33) constructs $X^\flat$ explicitly, but (6.34) defines $\alpha^\sharp$ only implicitly. Writing $\alpha = \beta_a dq^a + \gamma^a dp_a$ find the coordinate expression for $\alpha^\sharp$.

The symplectic 2-form can also provide us with an additional structure. If we wedge it with itself n times,

$$\Omega = \underbrace{\omega \wedge \cdots \wedge \omega}_{n \text{ copies}} = \pm n! \, dq^1 \wedge \cdots \wedge dq^n \wedge dp_1 \wedge \cdots \wedge dp_n, \tag{6.35}$$

we obtain a non-vanishing $2n$-form in a $2n$-dimensional space. Thus the symplectic structure endows $T^\star M$ with a measure, by which we can quantify volumes on $T^\star M$ without having to assume any additional metric structure.

6.4 HAMILTONIAN MECHANICS

We have now developed all the mathematical structures we need to describe classical mechanics on the cotangent bundle. We could simply postulate the basic laws *ab initio*. However, having already developed the formalism of Lagrangian mechanics on the tangent bundle, it is worth exploring to what extent this can be imported onto the cotangent bundle.

The Lagrangian itself can be used to construct a map from the tangent to the cotangent bundle, given by

$$(x^a, v^b) \mapsto \left(q^a = x^a, p_b = \frac{\partial L}{\partial v^b} \right). \tag{6.36}$$

$\{p_a\}$ here are the *canonical momenta*. In some cases, they correspond to the kinetic momentum given by mass times velocity[4], but this is not true in general. For example, the "standard" 2-dimensional Lagrangian in polar coordinates is

$$L = \frac{1}{2}m(v_r^2 + r^2 v_\theta^2) - \mathcal{V}(r, \theta). \tag{6.37}$$

The corresponding momenta are

$$p_r = mv_r, \quad p_\theta = mr^2 v_\theta. \tag{6.38}$$

p_r is the radial component of the kinetic momentum, but p_θ is the *angular momentum*.

Exercise 6.6: As we have seen, the Lagrangian for a charged particle in a magnetic field is $L = \frac{1}{2}mg_{ab}v^a v^b + qA_a v^a$, where A_a are components of the magnetic potential 1-form. In Cartesian coordinates, find the canonical momentum arising from this Lagrangian. How is this related to the kinetic momentum?

The map (6.36) can trivially be solved for $\{x^a\}$ in terms of $\{q^a, p_b\}$. Solving for $\{v^a\}$ is less straightforward, and not guaranteed to be possible. A Lagrangian that cannot be inverted to solve for velocities is called *singular*; we shall briefly consider singular Lagrangians in Section 6.7,

---

[4]Using some metric tensor to switch between vectors (velocities) and 1-forms (momenta).

but for the present, we will assume the Lagrangian is *regular* so that we can write the velocities in terms of positions and momenta.

With a slight abuse of notation, we now take L to be the function on T^*M which gives the same value as the Lagrangian on TM, using the inverse of the map (6.36). We then use its *Legendre transform* to define the Hamiltonian:

$$H = p_a v^a - L, \tag{6.39}$$

where the velocities $\{v^a\}$ are functions of $\{q^a, p_b\}$.

The action can now be expressed as a functional of curves on T^*M:

$$S[\Gamma] = \int_\Gamma (p_a \dot{q}^a - H)\, dt. \tag{6.40}$$

We do not assume here that Γ is the lift of some curve on M. However, we can construct a curve on M by projection: $\gamma(t) = \pi(\Gamma(t))$. The tangent to this curve is $\gamma' = \dot{q}^a \partial_a$, with $\{\dot{q}^a = dq^a/dt\}$ the evaluation of $\{v^a\}$ on Γ.

Just as on the tangent bundle, the action should be stationary for variations around physical paths:

$$0 = \delta S = \int_\Gamma \left(\delta p_a \dot{q}^a + p_a \delta \dot{q}^a - \frac{\partial H}{\partial q^a} \delta q^a - \frac{\partial H}{\partial p_a} \delta p_a \right) dt$$

$$= p_a \delta q^a \Big|_{\partial \Gamma} + \int_\Gamma \left[\left(\dot{q}^a - \frac{\partial H}{\partial p_a} \right) \delta p_a - \left(\dot{p}_a + \frac{\partial H}{\partial q^a} \right) \delta q^a \right] dt. \tag{6.41}$$

As we saw on the tangent bundle, the arbitrariness of the variations gives rise to equations of motion. However, there is a subtle but important difference: the variations $\{\delta p_a\}$ and $\{\delta q^a\}$ are entirely independent. The paths on TM that yielded the Euler-Lagrange equations were lifts of paths on M, whereas those yielding Hamilton's equations are arbitrary paths in T^*M.

For variations $\{\delta q^a\}$ that vanish on $\partial \Gamma$, the requirement of stationary action is equivalent to Hamilton's equations:

$$\dot{q}^a = \frac{\partial H}{\partial p_a}, \quad \dot{p}_a = -\frac{\partial H}{\partial q^a}. \tag{6.42}$$

This form of Hamilton's equations is all very well, but it somewhat obscures their intrinsic meaning. Exhibiting them in a more geometric form both provides further insight into their nature and allows us more flexibility when trying to solve them. This can be achieved by appealing to the symplectic structure of T^*M.

Since Hamilton's equations involve derivatives of the Hamiltonian, it is natural that the 1-form dH should be involved. In fact, the essential object is the Hamiltonian vector field, defined by

$$X_H = dH^\sharp = \frac{\partial H}{\partial p_a}\frac{\partial}{\partial q^a} - \frac{\partial H}{\partial q^a}\frac{\partial}{\partial p_a}. \tag{6.43}$$

Exercise 6.7: Verify that (6.43) satisfies $i_{X_H}\omega = dH$, where ω is the symplectic 2-form given in (6.29).

It immediately follows that (6.42) is equivalent to

$$\dot{z}^A = X_H z^A, \tag{6.44}$$

where $\{z^A\}$ are coordinates on $T^\star M$, for $A \in \{1, \ldots, 2n\}$. In fact, *any* function f on $T^\star M$ must be a function of the $\{z^A\}$, so it follows

$$\dot{f} = X_H f. \tag{6.45}$$

We can take this further. Observing that the right-hand side of (6.45) is the action on f of the Lie derivative with respect to X_H, we can use this to *define* the time derivative of a general tensor:

$$\dot{\Upsilon} = \mathcal{L}_{X_H}\Upsilon, \tag{6.46}$$

where Υ is *any* tensor field (scalar, vector, p-form, etc.) on $T^\star M$.

Using this notion of the Hamiltonian vector field generating evolution in time, we can explore the invariance (or otherwise) of the various structures we have considered.

The Hamiltonian can clearly be seen to be preserved:

$$\dot{H} = X_H H = i_{X_H} dH = i_{X_H} i_{X_H} \omega = 0, \tag{6.47}$$

the last equality following from the rules of the interior contraction. Likewise, the symplectic 2-form itself is preserved:

$$\begin{aligned}
\dot{\omega} &= \mathcal{L}_{X_H}\omega \\
&= i_{X_H}\underbrace{d\omega}_{=0} + d\underbrace{i_{X_H}\omega}_{=dH} \\
&= 0.
\end{aligned} \tag{6.48}$$

In the second line of (6.48) we have used Cartan's identity: when acting on differential forms, $\mathcal{L}_V = i_V d + d i_V$.

A further invariant that follows immediately from (6.48) is the phase space volume form:

$$
\begin{aligned}
\dot{\Omega} &= \mathcal{L}_{X_H} \Omega \\
&= \mathcal{L}_{X_H} (\underbrace{\omega \wedge \cdots \wedge \omega}_{n \text{ times}}) \\
&= n (\underbrace{\omega \wedge \cdots \wedge \omega}_{n-1 \text{ times}}) \wedge \mathcal{L}_{X_H} \omega \\
&= 0.
\end{aligned}
\tag{6.49}
$$

This last property ensures that Hamiltonian dynamics satisfies the requirements for Liouville's theorem, and hence phase space volume is preserved by the flow.

Exercise 6.8: Verify the invariance of H, ω and Ω by direct calculation for the harmonic oscillator, $H = p^2/(2m) + kq^2/2$.

6.5 POISSON BRACKETS

The Hamiltonian and symplectic 2-form together give rise to the Hamiltonian vector field. So far, we have seen this relation only indirectly:

$$
dH = i_{X_H} \omega.
$$

This is reminiscent of what we saw with the metric tensor, which could directly translate vectors into covectors, but only implicitly do the reverse. In that case, we defined the inverse metric tensor, with which we could directly generate the metric dual of a covector. We can do something similar with the symplectic 2-form.

We could define the inverse symplectic tensor. However, we are primarily interested in the vector field corresponding to an exact 1-form df. We therefore want an object which maps functions to vector fields. Moreover, since a vector field can be characterised by its action on scalar functions, we can focus on an object that maps two scalar fields to a single scalar field:

$$
\{f, g\} = dg^{\sharp} f \quad \text{or} \quad X_g = \{\cdot, g\}.
\tag{6.50}
$$

$\{f, g\}$ is called the *Poisson bracket* of the scalar field f with the scalar field g.[5]

The Poisson bracket is an intrinsic geometric object, but to make use of it, we need some way to represent it. In many cases, it is convenient to express the Poisson bracket in a *coordinate representation*. Generalising (6.43), we have

$$X_g = \frac{\partial g}{\partial p_a} \frac{\partial}{\partial q^a} - \frac{\partial g}{\partial q^a} \frac{\partial}{\partial p_a}, \tag{6.51}$$

and hence the Poisson bracket has the form

$$\{f, g\} = \frac{\partial f}{\partial q^a} \frac{\partial g}{\partial p_a} - \frac{\partial f}{\partial p_a} \frac{\partial g}{\partial q^a}. \tag{6.52}$$

In general phase space coordinates $\{z^A\}$, we can form a basis for the Poisson bracket,

$$\mathcal{I}^{AB} = \{z^A, z^B\}. \tag{6.53}$$

Then by the chain rule for derivatives, the Poisson brackets of any functions can be written

$$\{f, g\} = \frac{\partial f}{\partial z^A} \mathcal{I}^{AB} \frac{\partial g}{\partial z^B}. \tag{6.54}$$

There are a few key properties satisfied by the Poisson bracket. By construction, it is antisymmetric, and linear in both its arguments. It also satisfies the Leibniz product rule:

$$\{f, gh\} = \{f, g\}h + g\{f, h\}. \tag{6.55}$$

Finally, and less obviously, it satisfies the Jacobi identity:

$$\{f, \{g, h\}\} + \{g, \{h, f\}\} + \{h, \{f, g\}\} = 0. \tag{6.56}$$

The Jacobi identity is a consequence of the closedness of the symplectic 2-form.

Exercise 6.9: Using (6.54), write the Jacobi identity in terms of $\mathcal{I}^{AB}$ and its derivatives.

---

[5]If we require the symplectic dual of a non-exact 1-form, we can always use linearity to construct it from the Poisson bracket.

In terms of the Poisson bracket, Hamilton's equations take the simple form

$$\dot{f} = \{f, H\}, \tag{6.57}$$

with f an arbitrary phase space function and H the Hamiltonian. This form can be particularly effective, for example in the search for conserved quantities.

6.6 CONSERVED QUANTITIES

We have seen in our discussion of Lagrangian mechanics that there is a close relation between symmetries and conserved quantities. Further insight can be gleaned by exploring this relation in the context of symplectic geometry.

By definition, a conserved quantity g is a function on phase space that does not change in time. By (6.57), this implies its Poisson bracket with the Hamiltonian vanishes:

$$\frac{dg}{dt} = \{g, H\} = 0. \tag{6.58}$$

We expect then that the Poisson bracket of a phase space function is intimately connected with a corresponding transformation. This transformation will be a symmetry of the motion if its action on the Hamiltonian vanishes.

Consider that the Hamiltonian vector field X_H generates time translations:

$$X_H = \{\cdot, H\} = \frac{d}{dt} \tag{6.59}$$

(the second equality should be interpreted as indicating the action of the vector field on a scalar function). Similarly, any phase space function g gives rise to a vector field,

$$X_g = \{\cdot, g\}, \tag{6.60}$$

and this vector field generates some other transformation.[6] We may also say that the function g itself generates the transformation.

As an example, momentum p_a generates translations in the q^a direction:

$$X_{p_a} = \{\cdot, p_a\} = \frac{\partial p_a}{\partial p_b}\frac{\partial}{\partial q^b} - \frac{\partial p_a}{\partial q^b}\frac{\partial}{\partial p_b} = \frac{\partial}{\partial q^a}. \tag{6.61}$$

---

[6]Some texts refer to any vector field constructed in this way as a "Hamiltonian vector field", irrespective of whether g is the Hamiltonian function.

Hence if the Hamiltonian is invariant under translations in the q^a-direction, the corresponding momentum p_a is a conserved quantity.

Exercise 6.10: In $\mathbb{R}^3$, what transformation is generated by the z-component of angular momentum $xp_y - yp_x$?

From the antisymmetry of the Poisson bracket, we immediately find one conserved quantity: the Hamiltonian itself.[7]

Suppose a particular Hamiltonian H possesses two conserved quantities, f and g. From the Jacobi identity, we see that their Poisson bracket $\{f, g\}$ is also conserved:

$$\begin{aligned} \{\{f, g\}, H\} &= -\{\{g, H\}, f\} - \{\{H, f\}, g\} \\ &= -\{0, f\} - \{0, g\} \\ &= 0, \end{aligned} \tag{6.62}$$

where the first equality is a consequence of the Jacobi identity, and the second follows from the conservation of f and g. Note that this does not necessarily imply the existence of a new conserved quantity: $\{f, g\}$ may vanish, or be some function of f and g themselves.

Exercise 6.11: For the Hamiltonian $H = (p_x^2 + p_y^2)/2m$, the quantities $f = p_x$ and $g = yp_x - xp_y$ are conserved. Demonstrate explicitly that $\{f, g\}$ is also conserved.

Having explored some of the properties of conserved quantities, the question remains, why are they so important? The reader may already have a sense that these can facilitate the solution to the dynamics. This can be understood in terms of the orbits (i.e. solutions to Hamilton's equations) in T^*M.

The cotangent bundle is $2n$-dimensional, so in general to solve for the evolution of the system, we must solve $2n$ equations of motion. However, each constant of the motion places a restriction on where in phase space the orbit can lead. Thus, if we have r (independent) constants of the motion, the dynamics takes place on a $(2n-r)$-dimensional submanifold,

---

[7]Here we assume that the Hamiltonian is a function on T^*M, i.e. it does not depend on time. We will briefly consider how to handle the more general case in Section 6.7.

which may make it substantially easier to solve. Moreover, in numerical approximation schemes, it may be beneficial to work directly on the submanifold. This explicitly preserves the conserved quantities, which can prevent the approximate numerical evolution drifting far from the exact solution.

6.7 SINGULAR LAGRANGIANS

The transition from Lagrangian to Hamiltonian mechanics given in Section 6.4 relied on the possibility to invert the map $TM \to T^*M$, and hence to treat the velocities $\{v^a\}$ in terms of the $\{q^a, p_b\}$ variables. This is typically the case, but is certainly not guaranteed. In fact, there are a number of *singular Lagrangians* for which this inversion is not possible, including some very important examples.

If the Lagrangian is singular, it is still possible to derive a consistent Hamiltonian formulation. Rather than develop the theory in its entirety, we will illustrate the method with the example of a reparameterisation-invariant Lagrangian, and refer the interested reader to *Lectures on Quantum Mechanics* by Dirac.[8]

Consider the action

$$S = \int L(x, \dot{x}, t)dt. \tag{6.63}$$

Here we allow for the possibility that the Lagrangian depends explicitly on t. We also denote velocities as $\dot{x}$, since their relation to time parameterisation is crucial to the argument. For the same reason, we include the arguments of the Lagrangian, suppressing indices for conciseness (i.e. x implies $\{x^1, \ldots, x^n\}$, etc.).

We can rewrite this in terms of an arbitrary time parameter λ:

$$S = \int \tilde{L}(x, x', t, t')d\lambda, \quad \tilde{L}(x, x', t, t') = L(x, x'/t', t)t', \tag{6.64}$$

where a prime ($'$) denotes differentiation with respect to λ.

Although they give rise to the same action, the Lagrangians in (6.63) and (6.64) differ in two important respects. Firstly, while t is the independent parameter associated with L, it has become just another coordinate in $\tilde{L}$ (which has no explicit dependence on the new independent parameter λ). Secondly, while a change of parameterisation $t \to \tilde{t}(t)$ will

---

[8]Dirac, P. A. M. (2001). *Lectures on Quantum Mechanics*. New York: Dover Publications.

invalidate the form of (6.63), the equivalent change $\lambda \to \tilde{\lambda}(\lambda)$ leaves the form of (6.64) invariant.

Exercise 6.12: Show that setting $\lambda = t$ in (6.64) recovers (6.63).

Proceeding to the Hamiltonian formulation as usual, the momenta with respect to the new Lagrangian $\tilde{L}$ are given by

$$\tilde{p}_a := \frac{\partial \tilde{L}}{\partial x'^a} = p_a, \quad \tilde{p}_t := \frac{\partial \tilde{L}}{\partial t'} = \frac{1}{t'}\left(\tilde{L} - \tilde{p}_a x'^a\right). \tag{6.65}$$

It is clear from the arbitrariness in λ that it will not be possible to invert these relations to solve for the velocities (x', t'). However, we also see an even bigger problem: the new Hamiltonian

$$\tilde{H} := \tilde{p}_a q'^a + \tilde{p}_t t' - \tilde{L} = 0 \tag{6.66}$$

is seen to vanish identically.

We can overcome these difficulties by recalling that the relations in (6.65) do not *define* the momenta $(\tilde{p}_a, \tilde{p}_t)$, but rather are the outputs of the Lagrangian map from TM to T^*M, which now sends the entire dynamics to a submanifold of T^*M. In other words, the constraint (6.66) is not an identity, but a relation satisfied by physical solutions.[9]

Inspection of (6.65) shows that the quantity

$$\phi := \tilde{p}_t + H(q, \tilde{p}) \approx 0, \tag{6.67}$$

where H is the Hamiltonian corresponding to (6.63), and "$\approx$" denotes equality on the space of physical solutions (this type of equality is called *weak equality*).

It follows that we can write

$$\tilde{H} \approx \tilde{H}_T := v\phi, \tag{6.68}$$

where $\tilde{H}_T$ is the "total Hamiltonian" and v is arbitrary, corresponding to the arbitrariness in the time parameter λ (v should not be mistaken for a velocity).

We can now return to the usual Hamiltonian equations using $\tilde{H}_T$,

$$f' \approx \{f, \tilde{H}_T\} = \{f, v\}\phi + v\{f, \phi\} \approx v\{f, \phi\}, \tag{6.69}$$

---

[9]This is similar to our discussion of conserved quantities in Section 6.6, with the difference that, here, all solutions have the same value ($\tilde{H} = 0$) of the Hamiltonian.

making sure not to enforce the constraint (6.67) until after evaluating the Poisson brackets.

To illustrate the above ideas, we turn to the prototypical example of the harmonic oscillator. The parameterisation invariant Lagrangian for the harmonic oscillator is

$$\tilde{L} = \frac{mx'^2}{2t'} - \frac{k}{2}x^2 t'. \tag{6.70}$$

Using the relation $q = x$, the resulting momenta,

$$\tilde{p} = m\frac{q'}{t'}, \quad \tilde{p}_t = -\frac{1}{2}\left(m\frac{q'^2}{t'^2} + kq^2\right), \tag{6.71}$$

clearly satisfy the relation

$$\phi := \tilde{p}_t + \frac{\tilde{p}^2}{2m} + \frac{k}{2}q^2 \approx 0, \tag{6.72}$$

while the natural Hamiltonian

$$\tilde{H} = \tilde{p}q' + \tilde{p}_t t' - \tilde{L} = 0 \tag{6.73}$$

vanishes identically.

Taking instead the total Hamiltonian

$$\tilde{H}_T = v\phi, \tag{6.74}$$

we find Hamilton's equations become

$$q' = \{q, \tilde{H}_T\} \approx v\{q, \phi\} = v\frac{\tilde{p}}{m}, \tag{6.75}$$

$$t' = \{t, \tilde{H}_T\} \approx v\{t, \phi\} = v, \tag{6.76}$$

$$\tilde{p}' = \{\tilde{p}, \tilde{H}_T\} \approx v\{p, \phi\} = -vkq, \tag{6.77}$$

$$\tilde{p}_t' = \{\tilde{p}, \tilde{H}_T\} \approx v\{\tilde{p}, \phi\} = 0. \tag{6.78}$$

Dividing (6.75) and (6.77) by (6.76) yields the usual Hamilton equations for the harmonic oscillator parameterised by time, whilst (6.78) demonstrates the constancy of the standard Hamiltonian.

Exercise 6.13: The Lagrangian for a free relativistic particle can be written in the form $\tilde{L} = -m\sqrt{-\eta_{ab}\dot{x}^a\dot{x}^b}$, where η_{ab} is the lift of the components of the Minkowski metric tensor. Show that this Lagrangian is already invariant under reparameterisations, and that the natural Hamiltonian vanishes. Find a constraint surface $\phi \approx 0$, and write down the total Hamiltonian $\tilde{H}_T = v\phi$. From Hamilton's equations with $\tilde{H}_T$, determine what values of v correspond to (i) proper time parameterisation (i.e. $\eta_{ab}\dot{x}^a\dot{x}^b = -1$), and (ii) lab time parameterisation (i.e. $\dot{x}^0 = 1$).

In this section, we have explored the smallest tip of the very elegant iceberg that is Dirac's theory of constrained Hamiltonian systems. The reader who wishes to explore further would do well to refer to Dirac's monograph on the subject, *Lectures on Quantum Mechanics.*[10]

6.8 ALMOST SYMPLECTIC GEOMETRY

To conclude our discussion of dynamics on the cotangent bundle, we consider a situation that cannot be treated in Hamiltonian (or Lagrangian) mechanics: the motion of a point charge in the presence of magnetically charged matter.

As we have seen, Maxwell's equations forbid the existence of magnetic charges. This restriction can be relaxed by introducing the magnetic charge density 3-form ρ_M, in terms of which Gauss's law for magnetic charge becomes

$$dB = \rho_M. \tag{6.79}$$

However, this presents a problem for the Hamiltonian formulation, which makes use of the potential 1-form A. Setting $B = dA$ immediately implies $\rho_M = 0$ in (6.79), since $\rho_M = ddA = 0$. So the natural question is, can we construct a Hamiltonian formulation that does not rely on the potential 1-form?

Consider the Hamiltonian of a point charge in a magnetic field:

$$H = \frac{g^{ab}\left[p_a - qA_a\right]\left[p_b - qA_b\right]}{2m}. \tag{6.80}$$

It has been convenient to separate the coordinates on $T^{\star}M$ into those representing the point in configuration space $\{q^a\}$ and those representing the cotangent vector $\{p_a\}$. However, this is a choice, and we should not feel bound by it: we are free to choose any coordinates on $T^{\star}M$ that may be convenient.

Let us now define new coordinates (y^a, π_b) by

$$y^a = q^a, \qquad \pi_b = p_b - qA_b. \tag{6.81}$$

In terms of these coordinates, the Hamiltonian now reads

$$H = \frac{g^{ab}\pi_a\pi_b}{2m}, \tag{6.82}$$

[10]Dirac, P. A. M. (2001). *Lectures on Quantum Mechanics.* New York: Dover Publications.

which no longer contains the magnetic potential 1-form. The effects of the magnetic interaction now appear instead in the symplectic 2-form:

$$\omega = dy^a \wedge d\pi_a - qB \tag{6.83}$$

where B is (the lift of) the magnetic field 2-form. Crucially, it is not the potential but the magnetic field B that appears in (6.83).

Exercise 6.14: What is the Poisson bracket $\{f, g\}$ in terms of coordinates $\{y^a, \pi_b\}$? Use either its expression in terms of coordinates $\{q^a, p_b\}$, or its definition as the "inverse of the symplectic 2-form".

Having eliminated the need for the potential, we are now free to generalise to the case where B satisfies (6.79). However, we need to check in this case that (6.83) still satisfies the conditions for a symplectic 2-form.

To formulate Hamilton's equations, it is essential that we can construct a Hamiltonian vector field, which requires ω to be non-degenerate. The simplest way to demonstrate this is by wedging together three copies of ω:

$$\omega \wedge \omega \wedge \omega = -6dy^1 \wedge dy^2 \wedge dy^3 \wedge d\pi_1 \wedge d\pi_2 \wedge d\pi_3 \neq 0. \tag{6.84}$$

This implies that ω can be used to convert 1-forms into vectors and vice versa, without introducing any ambiguity. In particular, using (6.82), the Hamiltonian vector field is

$$X_H = dH^\sharp = \frac{\pi^a}{m}\frac{\partial}{\partial y^a} + \frac{q}{m}B_{ab}\pi^b\frac{\partial}{\partial \pi_a}, \tag{6.85}$$

where $B = \frac{1}{2}B_{ab}dq^a \wedge dq^b$ and $\pi^a = g^{ab}\pi_b$.

The other requirement of a symplectic 2-form is that it should be closed. In fact, for $\rho_M \neq 0$, ω *does not* satisfy this requirement:

$$d\omega = -qdB = -q\rho_M. \tag{6.86}$$

A structure satisfying all the conditions of a symplectic structure but for which $d\omega \neq 0$ is called an "almost symplectic structure". This does not obstruct the formulation of a consistent dynamics, but it does introduce a few differences from a truly symplectic structure, and care must be taken.

Exercise 6.15: The fact that the almost symplectic 2-form is not closed gives rise to a violation of the Jacobi identity by the Poisson bracket. This modified bracket is known as a "twisted Poisson structure". For the Poisson bracket determined in Exercise 6.14, calculate the "Jacobiator" $\{f, g, h\} = \{f, \{g, h\}\} + \{g, \{h, f\}\} + \{h, \{f, g\}\}$.

Unlike in the true symplectic case, the flow of (6.85) does not preserve the almost symplectic 2-form:

$$\mathcal{L}_{X_H}\omega = (di_{X_H} + i_{X_H}d)\omega,$$
$$= ddH + i_{X_H}(-q\rho_M),$$
$$= -\frac{q}{m}\pi^a i_{\partial_{y^a}}\rho_M. \tag{6.87}$$

Nevertheless, it does preserve the phase space volume $\Omega = \omega \wedge \omega \wedge \omega$,

$$\mathcal{L}_{X_H}\Omega = di_{X_H}\Omega,$$
$$= \frac{1}{m}d\left[\pi^a i_{\partial_{y^a}}\Omega\right] + \frac{q}{m}d\left[B_{ab}\pi^b i_{\partial_{\pi_a}}\Omega\right],$$
$$= 0, \tag{6.88}$$

where the first term in square brackets is closed because it contains a factor $d\pi^1 \wedge d\pi^2 \wedge d\pi^3$ and no dependence on y, and the second is closed because of the antisymmetry of B_{ab}. Therefore, the dynamics of charged particles in the presence of magnetic matter satisfies the conditions for Liouville's theorem. This is not always true for almost symplectic dynamics.

Magnetically charged matter is not the only application of almost symplectic geometry in physics. We end this section with an additional example, as an exercise for the reader.

Exercise 6.16: Consider the Hamiltonian $H = p_t + p^2/(2m) + kq^2/2$ for the harmonic oscillator in extended phase space with coordinates (q, t, p, p_t), but with the almost symplectic 2-form $\omega = dq \wedge dp + dt \wedge dp_t + \lambda p dq \wedge dt$, with λ a constant. Find $d\omega$ and the Hamiltonian vector field $X_H = dH^\sharp$, and calculate $\mathcal{L}_{X_H}\omega$ and $\mathcal{L}_{X_H}\Omega$. What is the physical meaning of λ?

Connections

7.1 INTRODUCTION

In Chapter 5, we saw how differential forms provide a powerful and elegant description of electrodynamics. One aspect we did not discuss was how charged particles respond to electromagnetic fields. To understand how a particle responds to a force, we need to describe its acceleration. This requires differentiating its velocity vector field "along itself". To do this, we need to introduce a new geometric object: a *connection*.

Why do we need an additional structure to differentiate vectors (and other tensors) in a particular direction? We have already seen that scalar functions can be differentiated along a curve by acting on them with the vector tangent to that curve. This is possible because scalar functions are defined directly on the manifold. More general tensors are not: they are defined on (products of) the tangent space T_pM and its dual, T_p^*M.

To differentiate tensors, then, we need to know how to associate vectors on the tangent space at a given point p with those on the tangent space at a different point p'. But this is not something intrinsic to the manifold; it requires a choice to be made. This is the purpose of a connection.

Exercise 7.1: In the Lie derivative, we have already encountered one way to differentiate tensors along the flow of a vector field. Give at least one reason why we cannot use $\mathcal{L}_V V$ for the acceleration of a velocity vector V.

DOI: 10.1201/9781003228943-7

7.2   COVARIANT DIFFERENTIATION

We will define the connection ∇ (pronounced "del" or "nabla")[1] through its action in covariant differentiation. The covariant derivative ∇_X along a vector X is a map from tensor fields to tensors of the same rank.[2] The covariant derivative is *tensorial* in the vector along which it differentiates, i.e.

$$\nabla_{fX+gY}T = f\nabla_X T + g\nabla_Y T, \qquad (7.1)$$

where f and g are any scalar functions, X and Y are vectors, and T is any tensor field.

The action of ∇_X on general tensor fields is built up from its action on 0-forms f,

$$\nabla_X f = Xf = df(X), \qquad (7.2)$$

and on a basis of vectors $\{X_a\}$,

$$\nabla_X X_a = \omega^b{}_a(X)X_b. \qquad (7.3)$$

Note that the structure of (7.3) follows directly from linearity in X, and the fact that the output must be a vector for each choice of a. The 1-forms $\{\omega^a{}_b\}$ are known as the *connection 1-forms* in the basis $\{X_a\}$; they completely define the connection. The components of $\omega^a{}_b$, given by

$$\Gamma^a{}_{bc} = \omega^a{}_c(X_b), \qquad (7.4)$$

are known as the *connection coefficients*. It is important to note that the connection coefficients are *not* the components of some tensor: the connection is a new kind of geometric object.

To complete the definition of ∇_X, we require it to be linear under addition,

$$\nabla_X(T + U) = \nabla_X T + \nabla_X U, \qquad (7.5)$$

and to satisfy the Leibniz rule for tensor products,

$$\nabla_X(T \otimes U) = (\nabla_X T) \otimes U + T \otimes (\nabla_X U). \qquad (7.6)$$

---

[1]Despite the notational similarity, ∇ is entirely distinct from the gradient operator ∇ of vector calculus. To avoid confusion, in this chapter we will always explicitly indicate when we are using the gradient operator.

[2]Note this distinction between tensor *fields* and tensors defined at a single point. If X is itself a vector field, the covariant derivative yields a tensor field. Except where we wish to emphasise the difference, we will take all tensors to be fields.

Note that both ordinary scalar multiplication and the exterior product on p-forms are special cases of the tensor product. Finally, we take covariant differentiation to commute with contraction: for example, if α is a 1-form and V a vector field,

$$\nabla_X(\alpha(V)) = (\nabla_X\alpha)(V) + \alpha(\nabla_X V). \tag{7.7}$$

Together, these rules allow us to determine the covariant derivative of any tensor. Consider, for example, a vector field $V = V^a X_a$. We have

$$\begin{aligned}
\nabla_X V &= \nabla_X(V^a X_a) \\
&= (\nabla_X V^a)X_a + V^a \nabla_X X_a \\
&= (XV^a)X_a + V^a \omega^b{}_a(X)X_b \\
&= [dV^a(X) + \omega^a{}_b(X)V^b]X_a. \tag{7.8}
\end{aligned}$$

Exercise 7.2: As we did with basis vectors, define the covariant derivative of an element of the dual cobasis $\{e^a\}$ as $\nabla_X e^a = \Omega^a{}_b(X)e^b$. Using the defining relation $e^a(X_b) = \delta^a_b$, determine $\Omega^a{}_b$ in terms of $\omega^a{}_b$.

The rules for covariant differentiation also allow us to determine how the connection 1-forms transform under a change of basis $X'_a = \Lambda^b{}_a X_b$. Then we have

$$\begin{aligned}
\omega'^a{}_b(V)X'_a &= \nabla_V X'_b \\
&= \nabla_V(\Lambda^c{}_b X_c) \\
&= d\Lambda^c{}_b(V)X_c + \Lambda^c{}_b \omega^d{}_c(V)X_d \\
&= \left[d\Lambda^c{}_b(V)\overset{-1}{\Lambda}{}^a{}_c + \Lambda^c{}_b \omega^d{}_c(V)\overset{-1}{\Lambda}{}^a{}_d \right] X'_a, \tag{7.9}
\end{aligned}$$

where $\overset{-1}{\Lambda}$ is the inverse of Λ, such that $X_a = \overset{-1}{\Lambda}{}^b{}_a X'_b$. Noting that the vector V is arbitrary and the basis $\{X'_a\}$ comprises linearly independent vectors, this leads to

$$\omega'^a{}_b = \overset{-1}{\Lambda}{}^a{}_d \omega^d{}_c \Lambda^c{}_b + \overset{-1}{\Lambda}{}^a{}_c d\Lambda^c{}_b. \tag{7.10}$$

The second term in (7.10) clearly demonstrates the non-tensorial nature of the connection. Moreover, it shows the necessity of a connection: even if the connection 1-forms vanish in some frame, they do not do so in general.

7.3 PARALLEL TRANSPORT

In the previous section, we established some of the properties of a connection. However, we said little about the role it can serve. In this section, we elucidate how it provides us with a notion of parallelism, and hence can be used to determine whether, and quantify by how much, tensors change from place to place.

The reader may well question why a geometric object needs to be introduced to decide whether two vectors are parallel: it feels very natural that we should be able to take a vector from "here" to "there", and compare it with another. This is because of a very particular property of the space we live in (or rather, a good approximation to the space we live in): absolute parallelism. In ordinary Euclidean space, we can transport a vector from point p to point p', and the result is always the same. In general, though, this is not true: the resulting vector at p' depends on the path taken from p to p'.

Consider a curve running between points p and p':

$$\gamma : \mathbb{R} \to M,$$
$$\lambda \mapsto \gamma(\lambda), \tag{7.11}$$

with $\gamma(0) = p$ and $\gamma(\lambda_f) = p'$, for some $\lambda_f > 0$. Then a vector $V_0 \in T_p M$ can be extended to a vector field $V \in T_\gamma M$ defined along γ, by requiring its covariant derivative to vanish:

$$\nabla_{\dot\gamma} V = 0, \tag{7.12}$$

where $\dot\gamma$ is the tangent vector to γ.[3] The vector field V is the *parallel transport* of V_0 along γ.

Exercise 7.3: Show that parallel transport is preserved under reparameterisations of γ, i.e. if we replace λ with a new parameter $\tilde\lambda(\lambda)$, the same vector field is obtained.

To illustrate parallel transport in action, consider a 2-dimensional space with the coordinate frame $\{X_1 = \partial_r, X_2 = \partial_\theta\}$. In this frame, take the connection 1-forms to be $\omega^1{}_1 = f dr$, $\omega^1{}_2 = -r(1 + rf)d\theta$, $\omega^2{}_1 = r^{-1}(1 + rf)d\theta$, and $\omega^2{}_2 = r^{-1}(1 + rf)dr$, where f is an arbitrary

---

[3]We previously used γ' to denote the tangent to γ. However, it is more convenient here to use $\dot\gamma$.

function of r. Now consider a circular path γ parameterised by θ, such that $\dot\gamma = X_2$.

For a vector $V = V^1 X_1 + V^2 X_2$, the parallel transport equation (7.12) becomes

$$
\begin{aligned}
0 &= \nabla_{\partial_\theta}(V^1 X_1 + V^2 X_2) \\
&= \frac{dV^1}{d\theta} X_1 + V^1 \omega^b{}_1(\partial_\theta) X_b + \frac{dV^2}{d\theta} X_2 + V^2 \omega^b{}_2(\partial_\theta) X_b \\
&= \left(\frac{dV^1}{d\theta} - r(1 + rf)V^2 \right) X_1 + \left(\frac{dV^2}{d\theta} + r^{-1}(1 + rf)V^1 \right) X_2. \quad (7.13)
\end{aligned}
$$

Note that r is constant along γ, so the components $\{V^1, V^2\}$ are functions of θ only.[4]

Recognising that X_1 and X_2 are linearly independent, the 2 components of (7.13) must each vanish, and combine to give

$$
\frac{d^2 V^a}{d\theta^2} + (1 + rf)^2 V^a = 0, \tag{7.14}
$$

for $a \in \{1, 2\}$. This clearly has solutions of the form $V^a = V_0^a \cos[(1 + rf)\theta]$. Hence the components of V are periodic, but not with period 2π. In general, after a full rotation, the parallel transported vector has returned to its original position, but not its original value. This demonstrates that parallel transport is *path dependent*.

Exercise 7.4: The definition of parallel transport is clearly independent of basis. Repeat the above example, relative to the basis $\{X_1' = e^{-F}\partial_r, X_2' = e^{-F}r^{-1}\partial_\theta\}$, where F is a function of r satisfying $f = dF/dr$. You will need to use (7.10) to determine the connection 1-forms relative to the new basis.

A particularly important instance of parallel transport is when the tangent vector itself is parallel transported,

$$
\nabla_{\dot\gamma}\dot\gamma = 0. \tag{7.15}
$$

In this case, the tangents to the curve at neighbouring points are always parallel, so we call the curve an "autoparallel". Autoparallels are the natural generalisation of straight lines in Euclidean geometry.

---

[4] Along a more general curve, the components would still be functions of a single variable, but r and θ would then each be functions of that variable.

Recall that parallel transport is independent of parameterisation. However, the same is not true of (7.15): under a reparameterisation $\lambda \to \tilde{\lambda}$, the tangent becomes $\dot{\gamma} \to \gamma' = \alpha\dot{\gamma}$, with $\alpha = d\lambda/d\tilde{\lambda}$. Using the rules of covariant differentiation, the autoparallel equation then becomes

$$\nabla_{\gamma'}\gamma' - \alpha^{-1}\frac{d\alpha}{d\tilde{\lambda}}\gamma' = 0. \qquad (7.16)$$

In fact, any equation of the form $\nabla_{\gamma'}\gamma' = f\gamma'$ can be transformed into (7.15) by a suitable reparameterisation.

An autoparallel satisfying (7.15) is said to be *affinely parameterised*. In general, the physical and geometric interpretation of an affine parameter can be somewhat obscure. However, in many important cases, it does have a clear meaning. We will return to this point when we look at the relationship between a connection and the metric tensor in Section 7.5.2.

7.4   GEODESICS AND THE LEVI-CIVITA CONNECTION

It is a well-known result in Euclidean geometry that the shortest distance between two points is a straight line. A natural question is whether this remains true more generally. To address this question, we begin by determining the length of a curve $\gamma(\lambda)$ between two points p_1 and p_2:

$$L = \int_\alpha^\beta \sqrt{g(\dot{\gamma}, \dot{\gamma})}d\lambda, \qquad (7.17)$$

where g is a positive definite metric tensor, and $\dot{\gamma}$ is the tangent vector to γ, both understood to be evaluated at $\gamma(\lambda)$. The initial and final values in the integral are such that the curve extends from p_1 to p_2, i.e. $\gamma(\alpha) = p_1$ and $\gamma(\beta) = p_2$.

Exercise 7.5: The length between two points may depend on the path taken, but not on the parameter of the curve. Prove that reparametrising the curve $\lambda \mapsto \tilde{\lambda}(\lambda)$ leaves L unchanged.

The shortest path from p_1 to p_2 will be that for which L is stationary under variations of γ.[5] To determine conditions for this, it is convenient to work in a coordinate frame $\{\partial_a = \partial/\partial x^a\}$, in which $\dot{\gamma} = t^a\partial_a$ and

---

[5]The following argument is similar to that used to determine the Euler-Lagrange equations in Chapter 6.

$g = g_{ab}dx^a \otimes dx^b$:

$$L = \int_\alpha^\beta \sqrt{g_{ab}t^a t^b}\, d\lambda, \tag{7.18}$$

and the variation of path length is

$$\delta L = \int_\alpha^\beta \frac{1}{2\sqrt{g(\dot\gamma, \dot\gamma)}} \left[\delta g_{ab}t^a t^b + 2g_{ab}t^a \delta t^b \right] d\lambda. \tag{7.19}$$

Denoting the variation in the coordinates of the curve by ξ^a, the induced variations in the components of the metric and tangent vector are $\delta g_{ab} = \xi^c \partial_c g_{ab}$ and $\delta t^a = d\xi^a/d\lambda$, respectively, hence

$$\delta L = \int_\alpha^\beta \frac{1}{2\sqrt{g(\dot\gamma, \dot\gamma)}} \left[\xi^c \partial_c g_{ab}t^a t^b + 2g_{ac}t^a \frac{d\xi^c}{d\lambda} \right] d\lambda. \tag{7.20}$$

To isolate ξ^c in (7.20), we must integrate the last term by parts. Before doing so, it is convenient to take the parameter λ to be arc-length, such that $g(\dot\gamma, \dot\gamma) = 1$ (note that we can only do this *after* varying the curve, as λ cannot be assumed to be arc-length along the alternative curves). Then we have

$$\delta L = \int_\alpha^\beta \xi^c \left[\left(\frac{1}{2}\partial_c g_{ab} - \partial_b g_{ac} \right) t^a t^b - g_{ac}\frac{dt^a}{d\lambda} \right] d\lambda. \tag{7.21}$$

Note that there are no boundary terms from the integration by parts, since by construction, ξ^c vanishes at the end points (we are considering the shortest path between fixed points p_1 and p_2).

Since ξ^c is arbitrary, the term in brackets must vanish. This can be rearranged to give

$$\frac{dt^a}{d\lambda} + \Gamma^a{}_{bc}t^b t^c = 0, \tag{7.22}$$

with

$$\Gamma^a{}_{bc} = \frac{1}{2}g^{ad} \left[\partial_b g_{dc} + \partial_c g_{bd} - \partial_d g_{bc} \right], \tag{7.23}$$

where g^{ad} are the components of the inverse metric, and we have used the symmetry of $t^a t^b$ to split the second term in (7.21) in two. Recognising that $\dot\gamma(t^a) = dt^a/d\lambda$, (7.22) is the a-component of the autoparallel equation

$$\nabla_{\dot\gamma}\dot\gamma = 0, \tag{7.24}$$

where ∇ here is the *Levi-Civita connection*, which has the connection coefficients (7.23).

Returning to our original question, we find that the shortest distance between two points is a straight line, if we define "straight" as being parallel with respect to the Levi-Civita connection. We refer to autoparallels of the Levi-Civita connection as "geodesics".

Exercise 7.6: Suppose now g is a spacetime metric with signature $s = 2$, and $\gamma(\lambda)$ is a worldline on this spacetime. Repeat the above analysis to find the equation of motion satisfied by γ, if it is to represent a geodesic. What differences are there from the purely spatial case?

7.5 TORSION AND NONMETRICITY

We have stressed that the connection is not itself a tensor. Nevertheless, it is possible to construct two important tensors from it: torsion and non-metricity. Notably, these both vanish for the Levi-Civita connection, which is constructed from the metric tensor. A general connection can be completely determined by specifying the metric, torsion, and nonmetricity tensors.

7.5.1 Torsion

Given a connection ∇, the covariant derivative can be considered as a map that acts on two vector fields and produces a third:

$$\nabla : (X, Y) \to \nabla_X Y. \tag{7.25}$$

This map acts like a tensor in its first argument (X), but not in its second (Y). This is because it differentiates Y:

$$\nabla_X(fY) = f\nabla_X Y + (Xf)Y. \tag{7.26}$$

However, we can "correct" this by defining a new tensor from the covariant derivative:

$$T(X, Y) = \nabla_X Y - \nabla_Y X - [X, Y], \tag{7.27}$$

where $[X, Y] = XY - YX$ is the commutator of X and Y (recall it is also the Lie derivative of Y with respect to X). We call T the *torsion* tensor of ∇.

Exercise 7.7: Verify that the torsion is a genuine tensor, i.e. $T(X, fY) = fT(X, Y)$ for any scalar field f. To complete this verification, argue (without additional calculation) that $T(fX, Y) = fT(X, Y)$.

The torsion tensor measures the deviation of the antisymmetrised covariant derivative from the Lie derivative. It can be used to encode the density of *dislocation defects* in a crystal lattice, or gravitational couplings to fermionic matter, for example.

An important class of connections are those for which the torsion vanishes, $T(X, Y) = 0$ for all vectors X and Y. Such connections are, for obvious reasons, called "torsion-free". They are also sometimes known as "symmetric".

Exercise 7.8: Given a basis $\{X_a\}$, its *structure coefficients* $\{f_{ab}{}^c\}$ are defined by $[X_a, X_b] = f_{ab}{}^c X_c$. In terms of $f_{bc}{}^a$ and the connection coefficients $\Gamma^a{}_{bc}$, determine the components of the torsion tensor, i.e. $T^a{}_{bc}$ where $T(X_b, X_c) = T^a{}_{bc} X_a$. If the connection is torsion-free, what additional properties must the basis satisfy for the connection coefficients to be symmetric in the lower two indices, i.e. $\Gamma^a{}_{bc} = \Gamma^a{}_{cb}$?

7.5.2 Non-metricity

The torsion tensor is defined entirely in terms of the connection. If our manifold is also equipped with a metric tensor, we can construct the *non-metricity tensor*, which is the covariant derivative of the metric:

$$Q(X, Y, Z) = \nabla_X g(Y, Z), \qquad (7.28)$$

where X, Y, Z are arbitrary vectors. Note that we first take the covariant derivative of g with respect to X, and then apply the resulting tensor to Y and Z. Non-metricity plays a role in some extensions of Einstein's general theory of relativity, though these remain speculative.

To understand the interpretation of non-metricity, we return to the concept of parallel transport. If a vector V is parallel transported along a curve γ, we are essentially saying that it is the *same vector* at all points of γ.[6] We might expect, therefore, that its length should not change

---

[6]In this sense, the name "parallel transport" does not capture its full essence.

along γ. However, instead, we find

$$\frac{d[g(V,V)]}{d\lambda} = \nabla_{\dot\gamma}[g(V,V)],$$
$$= \nabla_{\dot\gamma}g(V,V) + 2g(V,\nabla_{\dot\gamma}V),$$
$$= Q(\dot\gamma,V,V), \tag{7.29}$$

where in the second line we have used the symmetry of the metric to turn $g(\nabla_{\dot\gamma}V,V)$ into a second copy of $g(V,\nabla_{\dot\gamma}V)$. It follows that, in general, a vector undergoing parallel transport has constant length only if the non-metricity vanishes, $Q = 0$.

This argument also clarifies the value of using an affine parameter, in instances where the non-metricity vanishes. Setting $V = \dot\gamma$ in (7.29), and using the autoparallel equation $\nabla_{\dot\gamma}\dot\gamma = f\dot\gamma$ for arbitrary parameter-isation, we find

$$\frac{d[g(\dot\gamma,\dot\gamma)]}{d\lambda} = \nabla_{\dot\gamma}[g(\dot\gamma,\dot\gamma)],$$
$$= \nabla_{\dot\gamma}g(\dot\gamma,\dot\gamma) + 2g(\dot\gamma,\nabla_{\dot\gamma}\dot\gamma),$$
$$= Q(\dot\gamma,\dot\gamma,\dot\gamma) + 2fg(\dot\gamma,\dot\gamma). \tag{7.30}$$

Setting $Q = 0$, then, it follows that affine parameters (for which $f = 0$) are those for which the magnitude of the tangent vector remains constant along the curve. In particular, proper time s is an affine parameter.

Exercise 7.9: As we have seen, the angle θ between two vectors V and W is defined by $\cos\theta = g(V,W)/\sqrt{g(V,V)g(W,W)}$. If V and W are parallel transported along a curve γ, show that θ may vary if $Q \neq 0$. Show further that if the non-metricity tensor has the form $Q = \alpha \otimes g$ (i.e. $Q(X,Y,Z) = \alpha(X)g(Y,Z)$ for some 1-form α), then θ is constant along γ.

It is often convenient to restrict our considerations to connections for which $Q = 0$. If this is the case, ∇ is said to be *metric compatible*. With the Levi-Civita connection we have already seen that, given any metric g, we can find a connection such that $Q = 0$. We can invert this concept, and ask for any given connection, is it always possible to find a metric $\hat{g}$ such that $\nabla_X\hat{g} = 0$ for all vectors X? In fact, the answer to this question is in general no. The notion of metric compatibility is a stronger constraint on the connection than it is on the metric.

We can nevertheless explore how a metric compatible connection gives rise to non-metricity if we consider an alternative metric. For example, consider a metric tensor g and its associated Levi-Civita connection ∇. But suppose instead we wish to make measurements relative to the conformally related metric $\hat{g} = \exp(f)g$, where f is some scalar field.[7] Relative to this new metric, the non-metricity tensor has the form

$$
\begin{aligned}
Q(X,Y,Z) &= \nabla_X[\exp(f)g](Y,Z) \\
&= (Xf)\exp(f)g(Y,Z) + \exp(f)[\nabla_X g](Y,Z) \\
&= (df \otimes \hat{g})(X,Y,Z),
\end{aligned} \tag{7.31}
$$

where we have used $\nabla_X g = 0$. Hence $Q = df \otimes \hat{g}$, and the connection now exhibits non-metricity of a very specific form.

7.5.3 Reconstructing the connection

There are many other tensors we can construct from the connection, either on its own or in combination with the metric tensor. Indeed, we will meet perhaps the most important example in the next section. However, it is worth pausing here to see how the metric, the torsion, and the non-metricity can combine to uniquely determine the action of the connection.

Recall from our initial discussion that the covariant derivative of any tensor can be determined if we know its action on a set of basis vectors. In addition, since the metric tensor is invertible, knowledge of $g(V,W)$ for *arbitrary* vector W entirely determines V. Thus if we can express $g(\nabla_V W, X)$ for an arbitrary triple of vectors (V,W,X), we can determine the action of ∇ on any tensor.

It is a straightforward (if tedious) task to show that exactly this expression can be written as

$$
\begin{aligned}
2g(\nabla_V W, X) =&\, V[g(W,X)] + W[g(X,V)] - X[g(V,W)] \\
&+ Q(X,V,W) - Q(W,X,V) - Q(V,W,X) \\
&+ g(X,T(V,W)) + g(W,T(X,V)) - g(V,T(W,X)) \\
&+ g(X,[V,W]) + g(W,[X,V]) - g(V,[W,X]). \tag{7.32}
\end{aligned}
$$

---

[7] While this may seem somewhat artificial, two conformally related metrics are commonly employed in *scalar-tensor gravity* where, for example, g might relate to the "Einstein frame" and $\hat{g}$ to the "Jordan frame". Typically, the "Jordan frame" is the most natural from a mathematical perspective, whereas the "Einstein frame" is used to interpret the physical implications of the theory.

(Note the square brackets in the top line are used purely to indicate the order of operations, whilst those in the bottom line denote the Lie bracket, or commutator.) This identity can be verified by expanding $V[g(W, X)] = Q(V, W, X) + g(\nabla_V W, X) + g(W, \nabla_V X)$ and similar for the other terms in the top line, and using the definition of the torsion tensor.

While its proof is straightforward, the identity (7.32) may seem forbidding. The concerned reader should be reassured that there is rarely any call to use it in this fully general setting. We present it primarily to demonstrate that specifying the metric, torsion, and non-metricity tensors uniquely fixes the connection. Nevertheless, in many cases, (7.32) simplifies enormously. For example, if (V, W, X) are chosen to be elements of an orthonormal frame, the metric components are constant, and the first line of (7.32) does not contribute. Alternatively, if (V, W, X) are elements of a coordinate frame, their commutators vanish and the final line in (7.32) is zero.

Exercise 7.10: Show (7.32) reduces to the expression (7.23) for the Levi-Civita connection coefficients in a coordinate frame $\{\partial/\partial x^a\}$.

7.6 CURVATURE

Arguably the most important tensor we can construct from the connection is the Riemann curvature tensor. Indeed, many people will only knowingly encounter a connection in the context of a curved space, though as we have seen, it is meaningless to state that the connection 1-forms vanish, even in a flat space.

The most prominent appearance of the curvature tensor in physics is in Einstein's theory of gravity, the general theory of relativity. However, there are many other contexts in which it can play a role, for instance in continuum mechanics where it can model the density of *disclination defects* in a crystal lattice.

We begin by developing the mathematical form of the curvature tensor, in analogy with our construction of torsion. Since the covariant derivative of a vector is again a vector, we can evaluate the covariant derivative of the result, $\nabla_X \nabla_Y Z$.[8]

---

[8]We assume here that Y is a vector field, not merely a vector at a point.

The repeated covariant derivative $\nabla_X \nabla_Y Z$ is tensorial in its first argument, but not in either the second or third. However, as with the torsion, we can rectify this by "antisymmetrising" in X and Y, and including a term involving the commutator:

$$R(X,Y)Z = \nabla_X \nabla_Y Z - \nabla_Y \nabla_X Z - \nabla_{[X,Y]}Z. \qquad (7.33)$$

R is the famous *Riemann curvature tensor*. It depends on the connection, but in general not on the metric. If we specialise to the Levi-Civita connection, however, R does depend on the metric, since in this case ∇ itself is determined by the metric. Note that, in the present context, we interpret R as an (X,Y)-dependent linear transformation on Z.

Exercise 7.11: Verify that the Riemann curvature is a genuine tensor, i.e. $R(X,Y)(fZ) = fR(X,Y)Z$ and $R(X, fY)(Z) = fR(X,Y)Z$ for any scalar field f.

The Riemann curvature tensor contains a lot of information. It is therefore often convenient to consider alternative tensors, that distil some of the crucial aspects of R while discarding others.

One important example is the *Ricci tensor*. For given vectors Y and Z, the Ricci tensor is defined as the trace of the linear operator $X \mapsto R(X,Y)Z$, i.e.

$$\mathrm{Ric}(Y,Z) = e^a\Big(R(X_a,Y)Z\Big). \qquad (7.34)$$

Note that "Ric" should be interpreted as a single composite symbol (similarly to how the Reynolds number in fluid dynamics is denoted "Re"). Like the Riemann curvature, the Ricci tensor depends only on the connection, not the metric. However, its importance is best understood in the context of the Levi-Civita connection, in which case it describes the change in volume of a region as it is parallel transported. As such it plays an important role in the *singularity theorems* of black holes and the early universe.

Given a metric, the Ricci tensor can be further reduced to yield the *curvature scalar*,

$$\mathcal{R} = \mathrm{Ric}(\widetilde{e}^a, X_a). \qquad (7.35)$$

The curvature scalar is the simplest scalar that can be constructed from the fundamental geometrical objects, the connection and metric. Together with the Ricci tensor, it appears in Einstein's equations of general relativity,

$$\text{Ric} - \frac{1}{2}\mathcal{R}g = 8\pi GT, \tag{7.36}$$

where G is Newton's constant of universal gravitation, and the *stress-energy tensor* T is the source of the gravitational field.

There is one further object we can form from the Riemann curvature tensor. Sometimes known as the *Ricci 2-form*, ric(X,Y) is the trace of the linear map $Z \mapsto R(X,Y)Z$:

$$\text{ric}(X,Y) = e^a \Big(R(X,Y)X_a\Big). \tag{7.37}$$

This is much less widely used than Ric, but its value is in its obscurity. The 2-form ric encodes the failure of the connection to be compatible with *any* metric tensor. We will return to this once we have explored tensor-valued forms.

7.7 TENSOR-VALUED FORMS AND CARTAN'S STRUCTURE EQUATIONS

The antisymmetry of the torsion tensor in its two arguments is reminiscent of a 2-form. However, when a 2-form acts on a pair of vectors, the result is a scalar, whereas $T(X,Y)$ outputs a vector. Nevertheless, it would be convenient if we could harness the power of the exterior calculus of p-forms. We can do this by introducing the notion of *vector-valued forms*.

If we express the vector $T(X,Y)$ in terms of its components in a given frame,

$$T(X,Y) = 2T^a(X,Y)X_a, \tag{7.38}$$

then for each value of a, the quantity T^a is a 2-form, and we call the set $\{T^a\}$ a *vector-valued 2-form*. (The factor 2 is introduced in (7.38) for convenience, since our definitions imply that $i_Y i_X \beta = 2\beta(X,Y)$, for any 2-form β.) For many calculations, it is more convenient to work with the torsion 2-forms T^a than with the torsion tensor.

From (7.38) and the definition of the torsion tensor, we can express the torsion 2-forms through *Cartan's first structure equation*,

$$T^a = de^a + \omega^a{}_b \wedge e^b. \tag{7.39}$$

This is most readily proven in a coordinate frame, in which $de^a = 0$. Using the transformation rule (7.10) for the connection 1-forms, it can then be seen that (7.39) transforms as the components of a vector, and so holds in all frames.

Exercise 7.12: We can interpret the coframe e^a as a vector-valued 1-form. What is the corresponding tensor, i.e. if $\alpha(V) = e^a(V)X_a$, what is the mixed tensor corresponding to α?

We can generalise this notion to forms with more indices, but it is important to note that the connection 1-forms $\omega^a{}_b$ are *not* tensor-valued forms, since $\omega^a{}_b(X)$ do not form the components of any tensor.

The Riemann curvature tensor does provide us with a tensor-valued form. Since it is antisymmetric in its first two arguments, we can write

$$R(X,Y) = 2R^a{}_b(X,Y)e^b \otimes X_a, \qquad (7.40)$$

and for each choice of (a,b), $R^a{}_b$ is a 2-form.[9]

Similarly to what we saw with the torsion 2-forms, the curvature 2-forms can be elegantly expressed in terms of *Cartan's second structure equation*,

$$R^a{}_b = d\omega^a{}_b + \omega^a{}_c \wedge \omega^c{}_b. \qquad (7.41)$$

Exercise 7.13: Show that the curvature scalar (7.35) can be expressed as $\mathcal{R} \star 1 = R^a{}_b \wedge \star(\widetilde{X_a} \wedge e^b)$.

There is a third tensor-valued 1-form that is of interest to us here, the non-metricity 1-forms,

$$Q_{ab}(V) = Q(V, X_a, X_b). \qquad (7.42)$$

Although not going by the name of Cartan structure equations, the non-metricity 1-forms can be elegantly expressed as

$$Q_{ab} = dg_{ab} - (\omega_{ab} + \omega_{ba}), \qquad (7.43)$$

where g_{ab} are the components of the metric tensor, and $\omega_{ab} = g_{ac}\omega^c{}_b$. This can be derived by expressing $g = g_{ab} e^a \otimes e^b$, and expanding out the covariant derivative of g.

---

[9]The reader familiar with component-based tensor calculus should take care: $R^a{}_b$ are *not* the components of the Ricci tensor, but are 2-forms representing the full Riemann tensor.

We can now justify our claim that the Ricci 2-form describes the failure of the connection to be compatible with any metric. From the definition of the Ricci 2-form,

$$\mathrm{ric} = 2R^a{}_a = 2d\omega^a{}_a + 2\omega^a{}_b \wedge \omega^b{}_a. \qquad (7.44)$$

The second term vanishes by the antisymmetry of the exterior product, and the freedom to relabel repeated indices $(a, b) \leftrightarrow (b, a)$. From (7.43), then,

$$\mathrm{ric} = 2d\omega^a{}_a = dg^{ab} \wedge dg_{ab} - dQ, \qquad (7.45)$$

where $Q = g^{ab}Q_{ab}$ and g^{ab} are components of the inverse metric tensor. Since both ric and Q are frame independent, it follows that so too must be $dg^{ab} \wedge dg_{ab}$. In an orthonormal frame, the components of the metric tensor are constant, and hence $\mathrm{ric} = -dQ$. Since ric is defined without reference to any metric, then, a connection for which $\mathrm{ric} \neq 0$ exhibits non-metricity with respect to any metric.

Exercise 7.14: Use $g^{ac}g_{cb} = \delta^a_b$ to demonstrate $dg^{ab} \wedge dg_{ab} = 0$ directly, i.e. without appealing to properties of the connection.

7.8 FORCES AND ACCELERATION

An affinely parameterised curve γ satisfying $\nabla_{\dot\gamma}\dot\gamma = 0$ corresponds to a *straight line*. If γ is a timelike worldline, this means its *velocity is constant*, i.e. its acceleration vanishes. Clearly, then, $\nabla_{\dot\gamma}\dot\gamma$ is related to acceleration. In fact, if we take γ to be parameterised by proper time, $\nabla_{\dot\gamma}\dot\gamma$ is precisely the acceleration.

Through Newton's second law of motion, we can relate the acceleration to an applied force. However, since acceleration is a vector and force is typically described by a 1-form, this requires not only a connection ∇, but also a metric tensor g. In the following, we take ∇ to be the Levi-Civita connection associated with g, though it is possible to consider something more general.

7.8.1 Forces due to scalar fields

Some of the first forces to be described mathematically were *scalar forces*, e.g. Newton's law of universal gravitation and Coulomb's law of electrostatics. While these examples have been superseded by more sophisticated models, scalar forces remain relevant even in a relativistic setting, for example in describing the average response to a rapidly oscillating force.

The traditional, non-relativistic expression for the force on a particle in a scalar field Φ is

$$\mathbf{F} = -\boldsymbol{\nabla}\Phi, \tag{7.46}$$

where $\boldsymbol{\nabla}$ is the gradient operator (not to be confused with the connection). The natural translation of this into a geometric formulation would be

$$\mathfrak{F} = -d\Phi, \tag{7.47}$$

where $\mathfrak{F}$ is the force 1-form.

Using Newton's second law of motion with (7.47), we might expect a relativistic particle to satisfy

$$m\nabla_{\dot{\gamma}}\widetilde{\dot{\gamma}} = -d\Phi, \tag{7.48}$$

with ∇ the Levi-Civita connection. However, this leads to a problem. Consider how $g(\dot{\gamma}, \dot{\gamma})$ varies along γ:

$$\begin{aligned}
\nabla_{\dot{\gamma}}[g(\dot{\gamma}, \dot{\gamma})] &= Q(\dot{\gamma}, \dot{\gamma}, \dot{\gamma}) + 2g(\nabla_{\dot{\gamma}}\dot{\gamma}, \dot{\gamma}) \\
&= 2i_{\dot{\gamma}}\nabla_{\dot{\gamma}}\widetilde{\dot{\gamma}} \\
&= -\frac{2}{m}i_{\dot{\gamma}}d\Phi,
\end{aligned} \tag{7.49}$$

where we have used the vanishing of the non-metricity tensor, both in setting $Q(\dot{\gamma}, \dot{\gamma}, \dot{\gamma}) = 0$ and in writing $g(\nabla_{\dot{\gamma}}\dot{\gamma}, \dot{\gamma}) = i_{\dot{\gamma}}\nabla_{\dot{\gamma}}\widetilde{\dot{\gamma}}$. Since γ is parameterised by proper time, $g(\dot{\gamma}, \dot{\gamma}) = -1$, and the LHS of (7.49) should vanish. But the RHS is in general non-zero, so (7.48) cannot be correct.

One possible "fix" is to modify (7.47) by subtracting the component parallel to $\dot{\gamma}$. An alternative is to recognise that the short-hand "force equals mass times acceleration" is equivalent to Newton's second law only if the particle has a constant mass. The question of which approach is correct is largely a matter of interpretation, and both lead to the same result for the acceleration of $\dot{\gamma}$. Here, we follow the latter.

Newton's second law in full is $\nabla_{\dot{\gamma}}\mathfrak{p} = \mathfrak{F}$, where $\mathfrak{p} = m\widetilde{\dot{\gamma}}$ is the momentum. Expanding this out, we have

$$\dot{\gamma}(m)\widetilde{\dot{\gamma}} + m\nabla_{\dot{\gamma}}\widetilde{\dot{\gamma}} = -d\Phi, \tag{7.50}$$

where $\dot{\gamma}(m) = dm/d\lambda$ is the derivative of m along the worldline γ. Acting on (7.50) with $i_{\dot{\gamma}}$ (and recalling $g(\nabla_{\dot{\gamma}}\dot{\gamma}, \dot{\gamma}) = 0$) yields

$$\dot{\gamma}(m) = i_{\dot{\gamma}}d\Phi, \tag{7.51}$$

which reduces (7.50) to

$$m\nabla_{\dot\gamma}\widetilde{\dot\gamma} = -d\Phi - (i_{\dot\gamma}d\Phi)\,\widetilde{\dot\gamma} = i_{\dot\gamma}\left(\widetilde{\dot\gamma}\wedge d\Phi\right). \tag{7.52}$$

We now have an equation of motion that correctly describes the response of a particle to a scalar force. As anticipated, it is the same equation we would have got by subtracting the component of (7.47) parallel to $\widetilde{\dot\gamma}$. However, the variable mass makes it somewhat cumbersome to use. We can resolve this difficulty by noting from (7.51) that, along γ,

$$m = \Phi - \Phi_0, \tag{7.53}$$

where Φ_0 is a constant. Defining then the new scalar field $\Psi = \ln(\Phi/\Phi_0 - 1)$, the equation of motion (7.52) becomes

$$\nabla_{\dot\gamma}\widetilde{\dot\gamma} = i_{\dot\gamma}\left(\widetilde{\dot\gamma}\wedge d\Psi\right). \tag{7.54}$$

Here we have arrived at a consistent equation of motion with no reference to a varying mass.

As an alternative approach, we can describe the response to the scalar field Ψ in purely geometrical terms, taking the particle's worldline to be an autoparallel of a modified connection. Defining this new connection by its action on vector fields,

$$\bar\nabla_X Y = \nabla_X Y + (i_X d\Psi)Y - g(X,Y)\widetilde{d\Psi}, \tag{7.55}$$

it is immediately apparent that (7.54) is equivalent to

$$\bar\nabla_{\dot\gamma}\dot\gamma = 0. \tag{7.56}$$

Taking $Y = X_a$ in (7.55), we can generate the connection 1-forms

$$\bar\omega^a{}_b = \omega^a{}_b + \delta^a_b d\Psi - d\Psi(\tilde e^a)\tilde X_b, \tag{7.57}$$

from which it follows that $\bar\nabla$ is not compatible with the metric g. The non-metricity tensor is given by

$$\bar Q(X,Y,Z) = -2g(Y,Z)i_X d\Psi + g(Z,X)i_Y d\Psi + g(X,Y)i_Z d\Psi. \tag{7.58}$$

Autoparallels of a connection with non-metricity do not necessarily have tangent vectors with constant magnitude, so we must check that (7.56) preserves the normalisation condition:

$$\begin{aligned}
\bar\nabla_{\dot\gamma}[g(\dot\gamma,\dot\gamma)] &= Q(\dot\gamma,\dot\gamma,\dot\gamma) + 2g(\nabla_{\dot\gamma}\dot\gamma,\dot\gamma) \\
&= -2i_{\dot\gamma}d\Psi g(\dot\gamma,\dot\gamma) + i_{\dot\gamma}d\Psi g(\dot\gamma,\dot\gamma) + i_{\dot\gamma}d\Psi g(\dot\gamma,\dot\gamma) \\
&= 0. \tag{7.59}
\end{aligned}$$

Hence in this case, the autoparallels of $\bar{\nabla}$ are consistent with proper time parameterisation, so (7.56) is a valid equation of motion.

7.8.2 Electromagnetic forces

Arguably the most important force for describing the world around us is electromagnetism. The force exerted by electric and magnetic fields **E** and **B** on a point charge is described by the *Lorentz force*, which in classical vector calculus notation can be written

$$\mathbf{F} = q\left(\mathbf{E} + \mathbf{v} \times \mathbf{B}\right), \tag{7.60}$$

where q is the particle's charge and **v** its velocity.

A few key properties of (7.60) are immediately apparent, which suggest a natural way to restate it in terms of intrinsic geometrical concepts. Firstly, it is *linear* in the electric and magnetic fields, from which it follows that we want something linear in the electromagnetic 2-form F. Secondly, it makes use of the velocity, so the vector $\dot{\gamma}$ should also appear. Finally, since the force is represented by a 1-form, the contraction of F with $\dot{\gamma}$ has the correct structure. Hence we postulate that the Lorentz force 1-form is

$$\mathfrak{F} = q i_{\dot{\gamma}} F. \tag{7.61}$$

From our experience with the scalar force, we know it is important to check whether (7.61) gives rise to a consistent equation of motion, i.e. whether

$$m \nabla_{\dot{\gamma}} \dot{\gamma} = q \widetilde{i_{\dot{\gamma}} F} \tag{7.62}$$

preserves the normalisation condition $g(\dot{\gamma}, \dot{\gamma}) = -1$. We see

$$\begin{aligned}
\nabla_{\dot{\gamma}} \left[g(\dot{\gamma}, \dot{\gamma})\right] &= Q(\dot{\gamma}, \dot{\gamma}, \dot{\gamma}) + 2g(\nabla_{\dot{\gamma}} \dot{\gamma}, \dot{\gamma}) \\
&= 2\frac{q}{m} g(\widetilde{i_{\dot{\gamma}} F}, \dot{\gamma}) \\
&= 2\frac{q}{m} i_{\dot{\gamma}} i_{\dot{\gamma}} F \\
&= 0,
\end{aligned} \tag{7.63}$$

where we have used the vanishing non-metricity $Q = 0$, the definition of the metric dual $g(\tilde{\alpha}, V) = i_V \alpha$, and the antisymmetry of the interior contraction, $i_U i_V = -i_V i_U$. Unlike our naive first attempt at the scalar force 1-form, here we have an equation that is consistent with proper time normalisation.

As we saw in Chapter 5, an observer with velocity vector V perceives electric and magnetic fields $\mathcal{E} = i_V F$ and $\mathcal{B} = -i_V \star F$. An observer moving with the particle has $V = \dot\gamma$, and it follows that they interpret the force (7.61) as $\mathfrak{F} = q\widetilde{\mathcal{E}}$. Since such an observer would experience the particle as being at rest, this agrees with (the metric dual of) (7.60). We could verify that this remains true for an arbitrary observer. However, the intrinsic geometric nature of the Lorentz force means this is unnecessary: if it is true for some observer, it must be true for *all observers*.

Exercise 7.15: The 1-form $\mathfrak{F}' = \mu \star (F \wedge \widetilde{\dot\gamma})$ also satisfies the requirements of being linear in F and $\dot\gamma$, and preserves the proper time normalisation condition. However, it does not represent the Lorentz force on an electric charge. Suggest a possible interpretation for the force $\mathfrak{F}'$.

We can go some way towards solving (7.62). It will prove convenient to consider γ to be a member of a *congruence* of curves (i.e. a set of timelike worldlines such that exactly one passes through every point in a spacetime), all satisfying (7.62). Then their tangent vectors form a vector field V, satisfying

$$\nabla_V \widetilde{V} = \frac{q}{m} i_V F, \qquad g(V, V) = -1. \tag{7.64}$$

A useful identity for the Levi-Civita connection is $\nabla_X \widetilde{X} = i_X d\widetilde{X} + \frac{1}{2} d[g(X, X)]$.[10] Together with (7.64) this implies

$$i_V d\widetilde{V} = \frac{q}{m} i_V F. \tag{7.65}$$

From this, we readily infer

$$d\widetilde{V} = \frac{q}{m} F + \Omega, \tag{7.66}$$

where Ω is a 2-form satisfying $i_V \Omega = 0$ (to ensure agreement with the Lorentz force) and $d\Omega = 0$ (to ensure agreement with the Maxwell equation $dF = 0$).

In a 2-dimensional spacetime, the only 2-form satisfying these conditions is $\Omega = 0$. More generally, Cartan's identity implies $\mathcal{L}_V \Omega = 0$, so

---

[10]This can be proven by equating the actions of ∇_X and $\mathcal{L}_X$ on the 0-form $\widetilde{X}(Y)$ for an arbitrary vector Y, using the Cartan identity $\mathcal{L}_X \widetilde{X} = i_X d\widetilde{X} + d i_X \widetilde{X}$, the torsion-free condition $\nabla_Y X = \nabla_X Y - \mathcal{L}_X Y$, and $\widetilde{X}(\nabla_Y X) = \frac{1}{2} i_Y d[g(X, X)]$ from metric compatibility.

if $\Omega = 0$ across the congruence at some instant, it must remain zero at future times.

Setting $\Omega = 0$, it follows that $A = m\widetilde{V}/q$ can be used as a potential for F. This is particularly useful if we wish to treat the congruence of worldlines as a source for the electromagnetic field, $j = -qn \star \widetilde{V}$, where n is the *proper density of charge carriers*.

7.9 THE FERMI-WALKER DERIVATIVE

7.9.1 Gyroscopes and the concept of rotation

Accurate sensors for detecting changes in orientation occupy a critical role in modern technology. Gyroscopes are vital for satellite navigation, as well as for numerous terrestrial applications, and the gravitational field plays a prominent role in their behaviour. Although a gyroscope rotates around its spin axis, in the absence of an external torque, the spin itself does not rotate. However, the concept of rotation (specifically, what it means to be *non-rotating*) is quite subtle.

The simplest model of a gyroscope in relativity consists of two ingredients: the gyroscope's worldline γ and its spin vector S. The spin vector sits in the space of tangent vectors orthogonal to the gyroscope's four-velocity, i.e.

$$g(\dot{\gamma}, S) = 0. \tag{7.67}$$

From a Newtonian perspective, due to conservation of angular momentum, the gyroscope's spin will remain constant in magnitude and direction. The same conclusion holds from a relativistic perspective when the gyroscope is at rest in Minkowski spacetime. In this case $\dot{\gamma} = \partial/\partial t$, and the components of S with respect to $\partial/\partial x, \partial/\partial y, \partial/\partial z$ are constant along the gyroscope's worldline. Using (7.67), it can be seen that the $\partial/\partial t$ component of S is zero.

From a more general perspective, we expect S to be parallel transported along γ,

$$\nabla_{\dot{\gamma}}S = 0, \tag{7.68}$$

if the gyroscope is in free-fall, i.e. γ is a geodesic. Note that (7.68) is consistent with (7.67) because $\nabla_{\dot{\gamma}}(g(\dot{\gamma}, S)) = g(\nabla_{\dot{\gamma}}\dot{\gamma}, S) + g(\dot{\gamma}, \nabla_{\dot{\gamma}}S) = 0$ follows from the metric compatibility of the Levi-Civita connection ∇ and the geodesic equation

$$\nabla_{\dot{\gamma}}\dot{\gamma} = 0. \tag{7.69}$$

However, if the gyroscope is not in free-fall then, in general, (7.67) cannot hold if S is parallel transported along γ. Although the spin vector as perceived by an inertial observer would be constant, it would not be constant from the perspective of the gyroscope, i.e. from the perspective of an observer with the worldline γ. Ultimately, the spin is a physical property of the gyroscope, not the inertial observer, so any violation of (7.67) is physically unreasonable. The only viable option is to seek an alternative transport law. The way to proceed is to introduce the *Fermi-Walker derivative* $\nabla_{\dot\gamma}^F$ given by

$$\nabla_{\dot\gamma}^F X = \nabla_{\dot\gamma} X - \tilde{\mathcal{A}}(X)\,\dot\gamma + \tilde{\dot\gamma}(X)\,\mathcal{A}, \tag{7.70}$$

where $\mathcal{A} = \nabla_{\dot\gamma}\dot\gamma$ is the four-acceleration of the gyroscope, and require S to be *Fermi-Walker transported* along γ, i.e.

$$\nabla_{\dot\gamma}^F S = 0. \tag{7.71}$$

Note that Fermi-Walker transport coincides with parallel transport when γ is a geodesic, i.e. $\mathcal{A} = 0$.

Unlike the Levi-Civita connection ∇, the Fermi-Walker derivative $\nabla_{\dot\gamma}^F$ can only differentiate tensors along γ. So, although $\nabla_{\dot\gamma}^F$ is sometimes referred to as a connection in the literature, we will use the word "derivative" rather than "connection" to avoid confusion.

The general rules for the action of $\nabla_{\dot\gamma}^F$ on 0-forms and tensors other than vectors are similar to those of $\nabla_{\dot\gamma}$. In particular, $\nabla_{\dot\gamma}^F$ satisfies the Leibniz rule for tensor products, it commutes with contractions and satisfies $\nabla_{\dot\gamma}^F f = \dot\gamma(f)$ where f is any 0-form on the worldline γ.

Exercise 7.16: By applying $\nabla_{\dot\gamma}^F$ to the contraction $g(X, Y)$, where X and Y are vector fields, and by making use of (7.70), show $\nabla_{\dot\gamma}^F$ is compatible with the metric, i.e. $\nabla_{\dot\gamma}^F g = 0$.

Since $\nabla_{\dot\gamma}^F$ is compatible with the metric and it commutes with contractions, we have $\nabla_{\dot\gamma}^F (g(\dot\gamma, X)) = g(\nabla_{\dot\gamma}^F \dot\gamma, X) + g(\dot\gamma, \nabla_{\dot\gamma}^F X)$. However, inspection of (7.70) reveals $\nabla_{\dot\gamma}^F \dot\gamma = 0$, so $g(\dot\gamma, X)$ is constant if $\nabla_{\dot\gamma}^F X = 0$. Thus, as suggested above, (7.67) and (7.71) are indeed consistent.

One can introduce a triad X_1, X_2, X_3 of spacelike, unit normalised, mutually orthogonal, vectors that are orthogonal to $\dot\gamma$ and Fermi-Walker transported along γ. Thus, an observer with worldline γ perceives X_1, X_2, X_3 to be constant. The orthonormal frame $\{X_0 = \dot\gamma, X_1, X_2, X_3\}$

satisfies $\nabla^F_{\dot\gamma} X_a = 0$ and, by definition, is a *non-rotating* orthonormal basis for vectors over the worldline γ.

7.9.2 Classical behaviour of a particle with spin

The spin of a particle with a non-zero magnetic moment precesses in a magnetic field. If the particle is at rest then (using classical vector calculus notation) its spin vector **s** satisfies

$$\frac{d\mathbf{s}}{dt} = \boldsymbol{\mu} \times \mathbf{B} \tag{7.72}$$

where $\boldsymbol{\mu}$ is the magnetic moment of the particle. The magnetic moment is proportional to the spin, and is given by

$$\boldsymbol{\mu} = \frac{q}{2m}\mathfrak{g}\,\mathbf{s} \tag{7.73}$$

where, as before, q is the particle's charge and m is its mass. The dimensionless constant $\mathfrak{g}$ is the particle's g-factor.

Assuming $\mathbf{E} = 0$, inspection of the expression (5.15) for the electromagnetic 2-form F shows that the right-hand side of (7.72) can be written as $i_\mathrm{M}F$ where $\mathrm{M} = \mu_x\,\partial/\partial x + \mu_y\,\partial/\partial y + \mu_z\,\partial/\partial z$. Thus, introducing the particle's worldline γ, and noting $\dot\gamma = \partial/\partial t$ because the particle is at rest, we conclude that the particle's spin vector S satisfies $\nabla_{\dot\gamma}\widetilde{S} = i_\mathrm{M}F$.

The covariant generalisation of (7.72) when the particle's acceleration is non-zero has the form $\nabla^F_{\dot\gamma} S = \mathfrak{T}$, where the vector $\mathfrak{T}$ is the torque on the particle. However, the naive choice $i_\mathrm{M}F$ for the torque is incorrect. Since S is orthogonal to $\dot\gamma$, the properties of the Fermi-Walker derivative ensure the vector $\nabla^F_{\dot\gamma} S$ is orthogonal to $\dot\gamma$; thus, $\mathfrak{T}$ must be orthogonal to $\dot\gamma$. The simplest, and physically correct, approach is to repeat the strategy we used in Section 7.8.1 when discussing the force due to a scalar field. Introducing the expression

$$\widetilde{\mathfrak{T}} = -i_{\dot\gamma}(\widetilde{\dot\gamma} \wedge i_\mathrm{M}F) \tag{7.74}$$

for the metric dual of the torque, and making the substitution

$$\mathrm{M} = \frac{q}{2m}\mathfrak{g}\,S \tag{7.75}$$

yields the equation of motion

$$\nabla^F_{\dot\gamma}\widetilde{S} = -\frac{q}{2m}\mathfrak{g}\,i_{\dot\gamma}(\widetilde{\dot\gamma} \wedge i_S F) \tag{7.76}$$

for the spin of the particle. Note that, unlike $i_S F$, the right-hand side of (7.76) vanishes when contracted on $\dot{\gamma}$, even if $\mathbf{E} \neq 0$.

Exercise 7.17: Show (7.76) ensures that the magnitude of the spin vector S is constant along γ.

The previous exercise shows that, as required on physical grounds, an observer with worldline γ only perceives changes in the orientation of the particle's spin, not its magnitude.

Exercise 7.18: The simplest choice for the acceleration of the particle is the ratio of the Lorentz force and the mass:

$$\mathcal{A} = \frac{q}{m}\, i_{\dot{\gamma}}\widetilde{F}. \tag{7.77}$$

Use (7.77) to write (7.76) as the *Thomas-Bargmann-Michel-Telegdi* (TBMT) equation

$$\nabla_{\dot{\gamma}} S = \frac{q}{2m}\mathfrak{g}\, i_S\widetilde{F} + \frac{q}{2m}(2 - \mathfrak{g})\,(i_S i_{\dot{\gamma}} F)\,\dot{\gamma}. \tag{7.78}$$

Inspection of (7.78) shows $\mathfrak{g} = 2$ is a special case, and this is the value of the g-factor that emerges from the relativistic quantum mechanics of an electron.[11] Further considerations suggest corrections to (7.77) that couple the spin of the electron to the gradient of the magnetic field (so-called "Stern-Gerlach forces"). These additional terms induce corrections to the TBMT equation via the Fermi-Walker derivative in (7.76).

---

[11]Corrections to $\mathfrak{g} = 2$ of the order of the fine structure constant follow from quantum field theory.

CHAPTER 8

Generalised functions from a geometric perspective

8.1 INTRODUCTION

It is often helpful in physical calculations to make use of *generalised functions*. The 1-dimensional Dirac delta function $\delta(x)$ is perhaps the most famous example of a generalised function; it can be qualitatively thought of as an arbitrarily tall, but arbitrarily narrow, function that has unit area but vanishes at all points apart from $x = 0$. Consequently, given a continuous function $f(x)$, the Dirac delta function satisfies the identity $\int_a^b \delta(x) f(x)\, dx = f(0)$ if $a < 0 < b$. From a physical perspective, one can readily represent an impulsive force using the Dirac delta function. For example, using Newton's Second Law of Motion, the x-component $p(t)$ of the momentum of a particle subjected to an impulsive force at time $t = 0$ satisfies $dp/dt = p_0 \delta(t)$. The constant p_0 is the subsequent change in momentum since $\int_{t_1}^{t_2} \delta(t) dt = 1$, where $t_1 < 0 < t_2$, yields $p(t_2) - p(t_1) = \int_{t_1}^{t_2} (dp/dt) dt = p_0$.

As the notation suggests, $\delta(x)$ is commonly regarded in undergraduate physics simply as a function of x. However, mathematicians typically formulate the Dirac delta function in terms of a *linear map* on a space of *test functions* whose range is in the real line or the complex plane. There are different types of test function but we will only consider the simplest type: test functions that are *smooth* (i.e. their derivatives, to all orders, are continuous) and have *compact support*. The mathematically

DOI: 10.1201/9781003228943-8

precise meaning of the latter concept is outside the scope of this book; however, for our purposes, an intuitive grasp is sufficient. We simply state that functions with compact support vanish at all points outside a finite domain.[1] The linear maps on the test functions are known as *distributions* and, for clarity, we will subscript them with D. Furthermore, since each distribution is a functional (i.e. a function on a space of functions), square brackets are used to enclose its argument. For example, given a test function f, the distribution Δ_D corresponding to the Dirac delta function satisfies $\Delta_D[f] = f(0)$. As we will soon see, the more mathematical approach facilitates a natural, and computationally powerful, unification of exterior differential calculus with the calculus of integration.

8.2 THE EXTERIOR CALCULUS OF LINEAR DISTRIBUTIONS

It is straightforward to generalise the notion of distributions to the tensorial context. One merely requires the components of test tensors to be test functions (i.e. be smooth and have compact support). For our purposes, it is sufficient to focus on distributions that act on test *forms*, i.e. totally antisymmetric test tensors.[2] For example, let Σ be a 2-dimensional surface in a 4-dimensional spacetime M. The distribution Σ_D induced from the surface Σ is given by

$$\Sigma_D[\phi] = \int_\Sigma \phi \qquad (8.1)$$

where ϕ is a test 2-form. The fact that Σ_D is a linear map means

$$\Sigma_D[\lambda\,\phi + \mu\,\psi] = \lambda(\Sigma_D[\phi]) + \mu(\Sigma_D[\psi]) \qquad (8.2)$$

where μ, λ are constants, and ϕ, ψ are test 2-forms. Furthermore, (8.2) naturally generalises to more than one surface. Let Σ' be a second surface, and let Σ'_D be the corresponding distribution, i.e. $\Sigma'_D[\phi] = \int_{\Sigma'} \phi$. Then, by definition, $\lambda\,\Sigma_D + \mu\,\Sigma'_D$ is the distribution given by

$$(\lambda\,\Sigma_D + \mu\,\Sigma'_D)[\phi] = \lambda(\Sigma_D[\phi]) + \mu(\Sigma'_D[\phi]). \qquad (8.3)$$

---

[1]The *support* of a function is the region in which it is non-zero.

[2]There are several types of distribution in differential geometry. The specific type we will make use of here are commonly known as *de Rham currents*. However, since the emphasis here is on their role in physics, and not all de Rham currents describe physical currents, we will simply call them distributions. We prefer to reserve the term "current" for specific mathematical objects that describe physical currents (in particular, electric current).

In addition to inducing distributions in the above manner, one can also induce them from differential forms. For example, given a 2-form α in 4-dimensional spacetime, we have the distribution α_D given by

$$\alpha_D[\phi] = \int_M \alpha \wedge \phi \qquad (8.4)$$

where, by assumption, the test 2-form ϕ has compact support in the spacetime manifold M. Inspection of (8.4) shows α_D is a linear map. In particular, we have the result

$$\alpha_D[\lambda\,\phi + \mu\,\psi] = \lambda(\alpha_D[\phi]) + \mu(\alpha_D[\psi]) \qquad (8.5)$$

where μ, λ are constants, and ϕ, ψ are test 2-forms. Furthermore, it is straightforward to see

$$(\lambda\,\alpha_D + \mu\,\alpha_D')[\phi] = \lambda(\alpha_D[\phi]) + \mu(\alpha_D'[\phi]) \qquad (8.6)$$

where the distribution α_D' is induced from another 2-form α', i.e. $\alpha_D'[\phi] = \int_M \alpha' \wedge \phi$.

Equations (8.1) and (8.4) immediately unify the notion of the exterior derivative with the concept of the boundary of a region. To see this, note (8.1) yields

$$\begin{aligned}
\Sigma_D[d\chi] &= \int_\Sigma d\chi \\
&= \int_{\partial\Sigma} \chi \\
&= (\partial\Sigma)_D[\chi]
\end{aligned} \qquad (8.7)$$

where χ is a test 1-form and the generalised Stokes theorem (see Section 3.2.3) has been used in the second line. However, (8.4) leads to

$$\begin{aligned}
\alpha_D[d\chi] &= \int_M \alpha \wedge d\chi \\
&= \int_{\partial M} \alpha \wedge \chi - \int_M d\alpha \wedge \chi \\
&= -\int_M d\alpha \wedge \chi \\
&= -(d\alpha)_D[\chi].
\end{aligned} \qquad (8.8)$$

The second line in (8.8) follows from $d(\alpha \wedge \chi) = d\alpha \wedge \chi + \alpha \wedge d\chi$ (where the positive sign before the second term is obtained because α has even degree) and an application of the generalised Stokes theorem. The third line follows because the test 1-form χ has compact support in M, and so the first term in the second line is zero.

Drawing (8.7) and (8.8) together motivates the introduction of the following fundamental rules for the exterior calculus of distributions. Let p be the degree of the test forms on which a given distribution acts. The distribution is then said to have degree $q = n-p$ where n is the dimension of the underlying manifold (e.g. $n = 4$ if we are working in 4-dimensional spacetime). Furthermore, the exterior derivative of a q-*distribution* T_D is *defined* such that the rule

$$dT_D[\omega] = -(-1)^q T_D[d\omega] \qquad (8.9)$$

is obeyed.[3] The rule (8.9) is sometimes referred to as *integration by parts*, although it should be understood that this is just an analogy. Thus, (8.7) encourages the definition of the *boundary* ∂T_D of the distribution T_D via

$$\partial T_D[\omega] = T_D[d\omega]. \qquad (8.10)$$

Like "integration by parts", the word "boundary" is to be understood only as an analogy in this context. The identity

$$\partial T_D = (-1)^{q+1} dT_D, \qquad (8.11)$$

where T_D is a q-distribution, follows immediately from (8.9) and (8.10). Furthermore, the wedge product of a distribution T_D and any differential form β is defined to give the distribution $T_D \wedge \beta$ that satisfies[4]

$$(T_D \wedge \beta)[\omega] = T_D[\beta \wedge \omega]. \qquad (8.12)$$

However, we will not attempt to give meaning to the product of distributions as this would take us outside the realm of linear distribution theory. Non-linear distribution theory is harder, from a technical perspective, than the linear theory, and lies well outside the scope of this book. Fortunately, for our purposes, we have no need to multiply distributions. Although, using (8.12), a distribution can be wedged with

---

[3] What we describe here as a q-distribution is often called a p-*current* in the mathematics literature. For present purposes, it is more convenient to use a nomenclature that emphasises the integer q (i.e. $n - p$) rather than the integer p, although both conventions can be found elsewhere.

[4] We will not write, for example, $(T_D \wedge \beta)_D$, because we will only consider distributions whose products with smooth tensors are distributions. The additional D subscript is superfluous.

a differential form to give a new distribution, we will only consider the addition of distributions, not their multiplication.

The above properties lead directly to the distributional version

$$d(T_D \wedge \beta) = dT_D \wedge \beta + (-1)^q T_D \wedge d\beta \qquad (8.13)$$

of the graded Leibniz rule for the exterior derivative, where T_D is a q-distribution. To see this, we begin by noting

$$dT_D[\beta \wedge \omega] = -(-1)^q T_D[d(\beta \wedge \omega)] \qquad (8.14)$$

follows from (8.9). However, if the degree of β is p, we have

$$d(\beta \wedge \omega) = d\beta \wedge \omega + (-1)^p \beta \wedge d\omega. \qquad (8.15)$$

Thus, using the fact that T_D is a linear map, we can expand the right-hand side of (8.14) as

$$dT_D[\beta \wedge \omega] = -(-1)^q T_D[d\beta \wedge \omega] - (-1)^{p+q} T_D[\beta \wedge d\omega]. \qquad (8.16)$$

Using (8.12) to move the first argument of the wedge product in each term out of the square brackets in (8.16), and moving the first term from the right-hand side of the equation to the left-hand side, yields

$$(dT_D \wedge \beta + (-1)^q T_D \wedge d\beta)[\omega] = -(-1)^{p+q}(T_D \wedge \beta)[d\omega]. \qquad (8.17)$$

However, the distribution $T_D \wedge \beta$ has degree $p + q$ and so, using (8.9), we can write the right-hand side of (8.17) as $d(T_D \wedge \beta)[\omega]$ to give

$$(dT_D \wedge \beta + (-1)^q T_D \wedge d\beta)[\omega] = d(T_D \wedge \beta)[\omega]. \qquad (8.18)$$

Note that the argument in each term (the test form) is identical. The result (8.18) holds for all choices of test form ω, and so we can remove ω. We obtain the identity (8.13) as previously asserted.

In addition to extending the exterior derivative to distributions, we can also induce the action of the interior product on them via the definition

$$i_Y T_D[\omega] = -(-1)^q T_D[i_Y \omega] \qquad (8.19)$$

where Y is any vector field. Like the exterior derivative, the distributional version

$$i_Y(T_D \wedge \beta) = i_Y T_D \wedge \beta + (-1)^q T_D \wedge i_Y \beta \qquad (8.20)$$

of the graded Leibniz rule for the interior product follows as a consequence of (8.19) and (8.12).

Finally, the linearity of the distributional calculus allows us to expand the q-distribution T_D in terms of a basis of q-forms. In particular, given the basis $\{e^a\}$ for 1-forms, we have

$$T_D = \frac{1}{q!} T_{D\,a...b}\, e^a \wedge \cdots \wedge e^b \qquad (8.21)$$

where $\{T_{D\,a...b}\}$ are the components of T_D associated with the q-form basis $\{e^a \wedge \cdots \wedge e^b\}$. Note that each component of T_D is a 0-distribution because, using (8.12), we have

$$T_D[\omega] = \frac{1}{q!} T_{D\,a...b}[e^a \wedge \cdots \wedge e^b \wedge \omega] \qquad (8.22)$$

and the content of the square brackets is a test form of degree $q+p = n$, where n is the dimension of the underlying manifold.

Exercise 8.1: Like (8.9) and (8.12), the fundamental rule (8.19) for the interior product on distributions is modelled on the case where the distribution is induced from a differential form. Show (8.19) is compatible with

$$i_Y \alpha_D[\omega] = \int_M i_Y \alpha \wedge \omega \qquad (8.23)$$

where α is a q-form and ω is a test p-form, with $q = n - p + 1$.

8.2.1 The relationship with the Dirac delta function

Before turning to physical applications that showcase the above formalism, we will briefly consider its relationship with the physics-oriented notion of the Dirac delta function. As with the exterior calculus of differential forms, the most impressive physical applications of the exterior calculus of linear distributions arise in multi-dimensional contexts. However, to appreciate its relationship with more familiar methods, we will focus on the simplest case; that of test forms on an integration domain $\mathcal{I}$ equivalent to a finite interval in $\mathbb{R}$. We immediately have the distribution $\mathcal{I}_D$ given by

$$\mathcal{I}_D[\chi] = \int_{\mathcal{I}} \chi \qquad (8.24)$$

where χ is a test 1-form. However, if we introduce the coordinate x and express $\mathcal{I}$ as $x_1 \le x \le x_2$ we arrive at

$$\int_{\mathcal{I}} df = \int_{x_1}^{x_2} \frac{df}{dx} dx$$
$$= f(x_2) - f(x_1). \tag{8.25}$$

Thus, the application $\int_{\mathcal{I}} df = \int_{\partial \mathcal{I}} f$ of the generalised Stokes theorem yields the result $\partial \mathcal{I}_D[f] = f(x_2) - f(x_1)$.

Since $\mathcal{I}_D$ is a 0-distribution (which follows because $\mathcal{I}$ is equivalent to a 1-dimensional region (an interval) within a 1-dimensional space (the real line $\mathbb{R}$)), we see that $d\mathcal{I}_D$ is a 1-distribution. Thus, $\partial \mathcal{I}_D = -d\mathcal{I}_D$ immediately follows from (8.11) and we arrive at

$$d\mathcal{I}_D[f] = -f(x_2) + f(x_1). \tag{8.26}$$

Exercise 8.2: Verify (8.26) using (8.9) instead of the relationship between the boundary operator ∂ and exterior derivative d on distributions.

Inspection of (8.26) leads to the conclusion that $d\mathcal{I}_D$ is the sum of two distributions that resemble Dirac delta functions. To make this explicit, we introduce the 0-distribution $\delta_D^{x_0}$ given by

$$\delta_D^{x_0}[f dx] = f(x_0). \tag{8.27}$$

Note that an application of (8.12) leads to $(\delta_D^{x_0} dx)[f] = f(x_0)$ and so

$$d\mathcal{I}_D[f] = -(\delta_D^{x_2} dx)[f] + (\delta_D^{x_1} dx)[f] \tag{8.28}$$

immediately follows from (8.26). Finally, since f can be any test 0-form, we obtain

$$d\mathcal{I}_D = -\delta_D^{x_2} dx + \delta_D^{x_1} dx \tag{8.29}$$

which is immediately recognisable as a geometric version of the result

$$\frac{dI}{dx} = -\delta(x - x_2) + \delta(x - x_1) \tag{8.30}$$

where $I(x) = \theta(x_2 - x)\theta(x - x_1)$ is the top-hat function (with $\theta(x)$ denoting the Heaviside function) that vanishes outside the interval $x_1 < x < x_2$.

8.3 EDGE CURRENTS FROM A GEOMETRICAL PERSPECTIVE

The properties of 2-dimensional materials are a source of enduring interest in condensed matter physics. Much of the attention in recent years has focussed on so-called "topological" materials whose macroscopic properties are relatively robust to changes in the microscopic structure of the material or to changes in its environment. A famous and important example of their electronic behaviour arises in the context of the quantum Hall effect, where the Hall conductance of a sample of such a material immersed in a strong magnetic field exhibits an alternating sequence of jumps and plateaus when considered as a function of the magnetic field strength. The resistance is quantised in rational multiples of the von Klitzing constant, and this behaviour is so robust and universal it is now used by metrologists to define the ohm. Edge currents are a key aspect of the electronic behaviour of such materials and, as we will see, their macroscopic behaviour is highly amenable to analysis using the calculus of distributions.

Since the material is 2-dimensional, we should work with area-based (i.e. areal) densities instead of the usual volume-based densities. To avoid confusion, we will distinguish quantities that have area-based, rather than volume-based, units with a check. In particular, we introduce the electric charge *per unit area* $\check{\rho}$ and the electric current *per unit length* $\check{\mathbf{j}}$. We suppose that the sample of material is oriented so that it lies in the $z = 0$ plane. The electric current 2-form[5] is

$$j = -\check{\rho}\, dx \wedge dy + \check{j}_x dt \wedge dy - \check{j}_y dt \wedge dx \qquad (8.31)$$

where $\check{j}_x$, $\check{j}_y$ are the Cartesian components of $\check{\mathbf{j}}$. By assumption, the electric field has no component normal to the plane $z = 0$ and the magnetic field only has a normal component; thus, the Maxwell 2-form is

$$F = E_x\, dt \wedge dx + E_y\, dt \wedge dy - B_z\, dx \wedge dy. \qquad (8.32)$$

The electric current 2-form and Maxwell 2-form in the materials of interest here are related by

$$j = \check{\sigma} F \qquad (8.33)$$

---

[5]Note that j has units of charge, exactly like the electric current 3-form introduced in Chapter 5. Thus, although it is a slight abuse of notation, we will use the same symbol here. The reason for the minus sign in the third term in (8.31), when comparing j with the electric current 3-form, can be appreciated by showing $dj = 0$ agrees with the continuity equation in the $z = 0$ plane.

where the constant scalar $\breve{\sigma}$ is the *conductance*[6] of the material. Inspection of (8.31) and (8.32) shows that $\breve{j}_x = \breve{\sigma} E_y$ and $\breve{j}_y = -\breve{\sigma} E_x$; hence, $\breve{\mathbf{j}}$ is orthogonal to the electric field.[7]

Using the Gauss-Faraday law $dF = 0$, we immediately arrive at $dj = 0$ (i.e. the electric current is conserved). Whilst this result is reassuring, the 2-form j only describes part of the physical picture (the *bulk* electric current). As noted previously, edge currents play a prominent role in the quantum Hall effect and should be included in the mathematics.

Before turning to the details of the distributional generalisation of the electric current 2-form, it is worth noting that we do not need to invoke test forms that contain dz because the 2-form j does not contain dz. It is sufficient to work in the 3-dimensional "slice" of 4-dimensional spacetime at $z = 0$. From this perspective, the distribution $\mathcal{U}_D$ induced from the 3-dimensional region $\mathcal{U}$ swept out by the material sample has degree 0 (since the "slice" $z = 0$ and the region $\mathcal{U}$ are both 3-dimensional). To avoid over-complicating the calculation, we will regard the sample as "eternal"; i.e. the initial state and final state of the sample are in the distant past and far future, respectively. This arrangement can be implemented with a finite $\mathcal{U}$ by only using test forms whose support does not intersect the end caps of $\mathcal{U}$. For further simplicity, we will initially assume that the sample does not contain holes (although we will return to this question towards the end of this section). Thus, only one component of the boundary $\partial\mathcal{U}$ intersects with the support of the test forms; the 2-dimensional region swept out in spacetime by the edge of the sample. See Figure 8.1.

The total current (including the bulk and edge contributions) is then described by the 2-distribution j_D given by

$$j_D = \mathcal{U}_D j + \partial\mathcal{U}_D \wedge \mathsf{j} \tag{8.34}$$

where the 1-form j encodes the current flowing around the edge of the sample. Note that there is no need to explicitly introduce the wedge product in the first term in (8.34) because the distribution $\mathcal{U}_D$ has degree

---

[6]Whilst it is not uncommon for theorists to refer to the quantity $\breve{\sigma}$ as an area-based "conductivity", we will not do so here to avoid confusion. Our motivation for denoting the conductance as $\breve{\sigma}$ is given by the fact that σ is a common symbol for (conventional volume-based) conductivity, and conductance has units of conductivity multiplied by length.

[7]In general, the conductance of a material is a tensor. The quantity $\breve{\sigma}$ is often called the *Hall conductance* because it relates $\breve{\mathbf{j}}$ with the electric field in the same manner as the off-diagonal components of the conductance tensor.

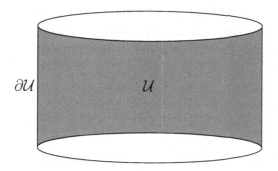

Figure 8.1 The 3-dimensional region $\mathcal{U}$ in spacetime swept out by a 2-dimensional circular sample. Time increases in the vertical direction. The boundary $\partial\mathcal{U}$ of $\mathcal{U}$ includes the 2-dimensional spacetime region swept out by the edge of the sample (the shaded surface), and the initial and final states of the sample (the end caps of $\mathcal{U}$). Only test forms that vanish on the end caps of $\mathcal{U}$ (the unshaded surfaces) are used.

0. Asserting the distributional version of electric current conservation,

$$dj_D = 0 \tag{8.35}$$

gives

$$0 = -\partial\mathcal{U}_D \wedge j - \partial\mathcal{U}_D \wedge dj \tag{8.36}$$

using the graded Leibniz rule (8.13), the fact that $\partial\mathcal{U}_D$ has degree 1, $dj = 0$, $d\mathcal{U}_D = -\partial\mathcal{U}_D$ and $d^2\mathcal{U}_D = 0$.

The meaning of (8.36) in more familiar terms is best illustrated using a simple example. Suppose that the sample of material is a disc of radius r_0. It is then natural to introduce plane polar coordinates r, θ via $x = r\cos\theta$, $y = r\sin\theta$ and express the 2-form (8.31) as

$$j = \check{\jmath}_r \, dt \wedge r_0 d\theta + \dots \tag{8.37}$$

where $\check{\jmath}_r = \check{\jmath}_x \cos\theta + \check{\jmath}_y \sin\theta$ and $\dots$ indicates terms that contain dr. Thus, the quantity $(\partial\mathcal{U}_D \wedge j)[f]$, where f is any test 0-form whose support does not intersect with the end caps of $\partial\mathcal{U}$, can be expressed as

$$(\partial\mathcal{U}_D \wedge j)[f] = \partial\mathcal{U}_D[jf]$$
$$= \int_{\partial\mathcal{U}} fj$$
$$= \int_{\partial\mathcal{U}} f\check{\jmath}_r \, dt \wedge r_0 d\theta. \tag{8.38}$$

The final step in (8.38) follows because dr is zero when restricted to any surface of constant r, such as the 2-dimensional region swept out by edge of the sample. This behaviour has an equivalent in the algebraic properties of $\partial \mathcal{U}_D$, as we will now see. The final line in (8.38) can be written as

$$\int_{\partial \mathcal{U}} f \check{j}_r \, dt \wedge r_0 d\theta = (\partial \mathcal{U}_D \wedge \check{j}_r \, dt \wedge r_0 d\theta)[f] \qquad (8.39)$$

and, since (8.38) holds for all choices of test 0-form f, we arrive at

$$\partial \mathcal{U}_D \wedge j = \partial \mathcal{U}_D \wedge \check{j}_r \, dt \wedge r_0 d\theta. \qquad (8.40)$$

Inspection of (8.40) suggests that the terms in j containing dr have been removed because of the wedge product, i.e. the distribution $\partial \mathcal{U}_D$ has the property $\partial \mathcal{U}_D \wedge dr = 0$.

Exercise 8.3: Show $\partial \mathcal{U}_D \wedge dr = 0$ by expressing $(\partial \mathcal{U}_D \wedge dr)[\chi]$ as an integral over $\partial \mathcal{U}$, where χ is any test 1-form whose support does not intersect with the end caps of $\partial \mathcal{U}$.

Since $\partial \mathcal{U}_D$ has degree 1, it must be possible to express it as a linear combination of the 1-form basis elements dt, dr, $d\theta$ with 0-distributions as coefficients. However, since $dt \wedge dr \neq 0$ and $d\theta \wedge dr \neq 0$, but $dr \wedge dr = 0$ (since dr has odd degree), we see $\partial \mathcal{U}_D \wedge dr = 0$ ensures that the coefficients of dt and $d\theta$ are zero. Thus, $\partial \mathcal{U}_D = \gamma_D dr$ where γ_D is a 0-distribution. Alternatively, one can make use of (8.20) to derive the same result.

Exercise 8.4: Use (8.20) and $\partial \mathcal{U}_D \wedge dr = 0$ to show $i_{\partial_t} \partial \mathcal{U}_D = 0$ and $i_{\partial_\theta} \partial \mathcal{U}_D = 0$, where ∂_t, ∂_θ are shorthand for $\partial/\partial t$, $\partial/\partial\theta$, respectively.

The fact $\partial \mathcal{U}_D = \gamma_D dr$ for some 0-distribution γ_D immediately allows (8.36) to be expressed entirely in terms of differential forms. Since the degree of γ_D is 0, it can be removed from (8.36) to give

$$dr \wedge dj = -dr \wedge j \qquad \text{at } r = r_0. \qquad (8.41)$$

Thus, expressing j as

$$j = -\check{\varrho} r_0 d\theta + \check{j}_\theta dt \qquad (8.42)$$

and introducing (8.37) reveals the componential version

$$\frac{\partial \check{\varrho}}{\partial t} + \frac{1}{r_0}\frac{\partial \check{\jmath}_\theta}{\partial \theta} = \check{\jmath}_r. \tag{8.43}$$

of (8.41), where $\check{\jmath}_\theta$ is the component of the edge current with respect to the unit vector in the direction of the angular coordinate θ around the circle at $r = r_0$. Accents have been included on the components of j as a reminder that they are line-based densities rather than area-based densities.[8]

The physical interpretation of (8.43) is clear. As a consequence of electric current conservation, the charge that flows out of the bulk of the material must join the charge flowing around the edge.

We will now show (8.41) readily reveals the relationship

$$\Delta Q = \check{\sigma}\Delta\Phi \tag{8.44}$$

between the change ΔQ in the charge flowing around the edge of the sample and the change $\Delta\Phi$ in the magnetic flux through the sample. Since (8.41) can be written as

$$d\mathsf{j} = -\check{\sigma} F + \dots \qquad \text{at } r = r_0 \tag{8.45}$$

where (8.33) has been used and $\dots$ indicates terms that contain dr, we immediately arrive at

$$\int_\Sigma d\mathsf{j} = -\check{\sigma}\int_\Sigma F \tag{8.46}$$

where Σ is any 2-dimensional region contained in the 2-dimensional surface swept out by the edge of the sample in spacetime. In particular, we choose Σ to be the cylinder whose ends terminate at $t = t_f$ and $t = t_i$.

The analysis of (8.46) requires a consideration of the relative orientations of the components of boundaries. Let $\mathcal{V}$ be the 3-dimensional region swept out by the sample between $t = t_i$ and $t = t_f$. The 2-dimensional boundary $\partial\mathcal{V}$ is the union of Σ and the 2-dimensional discs Σ_i, Σ_f that cap the cylinder Σ. In particular, we have

$$\int_{\partial\mathcal{V}} F = \int_{\Sigma_f} F - \int_{\Sigma_i} F + \int_\Sigma F \tag{8.47}$$

---

[8] The 1-form j, the 2-form j and the 2-distribution j_D all have physical units of charge. The 0-distribution $\mathcal{U}_D$ and 1-distribution $\partial\mathcal{U}_D$ are physically dimensionless.

where the minus sign has been introduced because, for simplicity, we have chosen the normal to both of the discs Σ_i, Σ_f to point in the direction of increasing t. However, $\int_{\partial V} F = \int_V dF = 0$ using the generalised Stokes theorem and the Gauss-Faraday law $dF = 0$. Hence

$$\int_\Sigma F = -\Phi_f + \Phi_i \qquad (8.48)$$

follows from introducing the magnetic fluxes $\Phi_f = \int_{\Sigma_f} F$, $\Phi_i = \int_{\Sigma_i} F$ through the sample at $t = t_f$, $t = t_i$, respectively.

Inspection of the orientation of the 2-dimensional surface Σ shows that the integral of j over the boundary $\partial \Sigma$ of Σ can be expressed as

$$\int_{\partial \Sigma} j = -\int_{C_f} j + \int_{C_i} j \qquad (8.49)$$

where $C_i = \partial \Sigma_i$, $C_f = \partial \Sigma_f$ are the closed curves that enclose the edge of the sample at the instants $t = t_i$, $t = t_f$, respectively. However, inspection of (8.42) shows $\int_{C_f} j = -Q_f$, where Q_f is the total charge of the edge current at $t = t_f$. The 1-form dt, and therefore the term $\check{j}_\theta dt$ in (8.42), vanishes when integrated over any curve of constant t. Likewise, $\int_{C_i} j = -Q_i$ where Q_i is the total charge of the edge current at $t = t_i$. Since $\int_\Sigma dj = \int_{\partial \Sigma} j$ follows from the generalised Stokes theorem, the result (8.44) is immediately obtained after introducing $\Delta Q = Q_f - Q_i$ and $\Delta \Phi = \Phi_f - \Phi_i$.

From a purely theoretical perspective, it is worth noting that the above discussion can be readily generalised to accommodate thought experiments involving fictitious samples whose shape is dynamical and, in general, non-planar. To see this, it is fruitful to move away from the 3-dimensional picture (time plus two spatial coordinates) introduced above and turn to a full 4-dimensional calculation. This is simpler than it may at first sound due to the succinct nature of the distributional calculus and the conceptual benefit of adopting a geometrical mindset.

The first step is to realise that the distribution $\mathcal{U}_D$ has degree 1 in the 4-dimensional context. The region $\mathcal{U}$ swept out by the sample in spacetime is 3-dimensional, and so $\mathcal{U}_D$ acts on test 3-forms; hence, the degree of $\mathcal{U}_D$ is $4 - 3 = 1$. Let ζ be a 0-form adapted to $\mathcal{U}$, i.e. $\mathcal{U}$ resides within a 3-dimensional region of constant ζ. The choice $\zeta = z$ corresponds to a planar sample but, in general, ζ can depend on t as well as all of the spatial coordinates. Since ζ is constant on $\mathcal{U}$ it follows $\int_\mathcal{U} d\zeta \wedge \phi = 0$ holds for all choices of test 2-form ϕ, and we conclude $\mathcal{U}_D \wedge d\zeta = 0$.

The region swept out by the edge of the sample can be represented as the 2-dimensional intersection of a 3-dimensional region of constant ξ with the region of constant ζ, where ξ is another 0-form. For example, the choices $\xi = r$ and $\zeta = z$ describe the disc-shaped sample discussed previously. As before, we regard the sample as "eternal", i.e. $\int_{\partial\mathcal{U}} d\xi \wedge \chi = 0$ and $\int_{\partial\mathcal{U}} d\zeta \wedge \chi = 0$ for all choices of test 1-form χ whose support does not intersect with the end caps of $\mathcal{U}$. Hence, it follows $\partial\mathcal{U}_D \wedge d\xi = 0$ and $\partial\mathcal{U}_D \wedge d\zeta = 0$.

Exercise 8.5: (a) Show that the distribution $\partial\mathcal{U}_D$ has degree 2.
(b) Show that $\partial\mathcal{U}_D$ can be written

$$\partial\mathcal{U}_D = \gamma_D d\xi \wedge d\zeta \tag{8.50}$$

where γ_D is a 0-distribution.

Hint: Begin part (b) by convincing yourself that it must be possible to express a general 2-distribution Θ_D as $\Theta_D = \gamma_D d\xi \wedge d\zeta + \mu_D \wedge d\xi + \nu_D \wedge d\zeta + \pi_D$ where the 1-distributions μ_D, ν_D and the 2-distribution π_D do not contain $d\xi$ or $d\zeta$.

The simplest way to generalise (8.34) is to assert that the electric current is described by the 3-distribution

$$j_D = \mathcal{U}_D \wedge \breve{\sigma} F + \partial\mathcal{U}_D \wedge \mathfrak{j} \tag{8.51}$$

instead of representing it using a distribution of degree 2 as done previously. Except for the fact that it does not contain $d\xi$ or $d\zeta$, the 1-form $\mathfrak{j}$ encoding the edge current is general.

The result $\partial\mathcal{U}_D = d\mathcal{U}_D$ follows from (8.10), because $\mathcal{U}_D$ has odd degree (it is a 1-distribution); hence the exterior derivative of (8.51) is

$$dj_D = \partial\mathcal{U}_D \wedge \breve{\sigma} F + \partial\mathcal{U}_D \wedge d\mathfrak{j} \tag{8.52}$$

since $dF = 0$ and $\breve{\sigma}$ is a constant. However, $dj_D = 0$ because electric charge is conserved; hence, inspection of (8.52) reveals

$$d\mathfrak{j} \simeq -\breve{\sigma} F \tag{8.53}$$

at the edge of the sample, where $\simeq$ indicates equality up to terms that contain one or both of $d\xi$ and $d\zeta$. The geometrical terminology for this type of equality is *equality after pullback* to $\partial \mathcal{U}$.[9] Notice that the structure of (8.53) is such that a hypothetical observer living entirely in the edge of the sample (i.e. an observer who cannot perceive spatial displacements away from the edge) would regard the term in $-\check{\sigma}F$ that survives the pullback to $\partial \mathcal{U}$ as a source for the edge current.

Integrating (8.53) over any 2-dimensional region Σ contained in the 2-dimensional region swept out by the edge of the sample in 4-dimensional spacetime yields

$$\int_{\Sigma} dj = -\check{\sigma} \int_{\Sigma} F \tag{8.54}$$

as before. Introducing the total charge $Q_t = -\int_{C_t} j$ in the edge C_t at time t, and the magnetic flux $\Phi_t = \int_{\Sigma_t} F$ through the sample Σ_t at time t, leads directly to (8.44) (where $\Delta Q = Q_f - Q_i$ and $\Delta \Phi = \Phi_f - \Phi_i$ as before) as a consequence of the generalised Stokes theorem.

Before closing this section, it is instructive to explore the behaviour of currents in samples that contain holes, i.e. samples with more than one edge. However, we need to take care with the relative orientations of the 2-dimensional regions swept out by the edges. From a distributional perspective, the natural approach is to assert the generalisation

$$j_D = \mathcal{U}_D \wedge \check{\sigma} F + \sum_{k=1}^{N} \Sigma_D^{(k)} \wedge j^{(k)} \tag{8.55}$$

of (8.51) to a sample with $N-1$ holes, where each k corresponds to one of the N edges. Each $\Sigma^{(k)}$ is the 2-dimensional region swept out by the k-th edge, and each edge has a current flowing around it described by the 1-form $j^{(k)}$.

Every 2-dimensional region $\Sigma^{(k)}$ in the summand in (8.55) has the same orientation, which must be borne in mind when considering the relative orientations of the components of the boundary $\partial \mathcal{U}$ of the 3-dimensional region $\mathcal{U}$ swept out by the sample. Note that, since the sample is "eternal" (we only make use of test forms whose support does

---

[9]The details are beyond the scope of this book, but it is perhaps worth commenting on the reason for this terminology. It turns out that a map from one manifold (in the present case, $\partial \mathcal{U}$) to another manifold (in the present case, 4-dimensional spacetime) induces a natural, geometrical, way of mapping (i.e. *pulling back*) differential forms from the second manifold to the first manifold. The induced map sends a subset of the differential forms on the second manifold to zero on the first manifold, leading to the type of equality described in the main text. The full details of this concept can be found in the suggestions for further reading given at the end of this book.

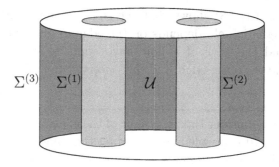

Figure 8.2 The 3-dimensional region $\mathcal{U}$ in spacetime swept out by a circular sample with two holes. Time increases in the vertical direction. Only test forms that vanish on the end caps of $\mathcal{U}$ (the unshaded surfaces) are used. The outer curved faces of the two cylinders swept out by the holes are denoted $\Sigma^{(1)}$ and $\Sigma^{(2)}$. The surface swept out by the outer edge of the sample is denoted $\Sigma^{(3)}$.

not intersect with the end caps of $\mathcal{U}$), each 2-distribution $\Sigma_D^{(k)}$ satisfies $\partial \Sigma_D^{(k)} = 0$. See Figure 8.2.

Identifying the outer edge of the sample with $k = N$ leads to

$$\partial \mathcal{U}_D = \Sigma_D^{(N)} - \sum_{k=1}^{N-1} \Sigma_D^{(k)} \tag{8.56}$$

where each $\Sigma^{(k)}$ for $k < N$ is the 2-dimensional region in spacetime swept out by the edge of a hole. Thus, the exterior derivative of (8.55) can be expressed as

$$dj_D = \Sigma_D^{(N)} \wedge \left(\check{\sigma} F + dj^{(N)} \right) + \sum_{k=1}^{N-1} \Sigma_D^{(k)} \wedge \left(- \check{\sigma} F + dj^{(k)} \right) \tag{8.57}$$

and the distributional representation of electric current conservation $(dj_D = 0)$ immediately reveals

$$dj^{(N)} \simeq -\check{\sigma} F,$$
$$dj^{(k)} \simeq \check{\sigma} F \quad \text{for } k < N \tag{8.58}$$

where $\simeq$ indicates equality after pullback to the 2-dimensional region labelled by the superscript on the edge current 1-form.

Inspection of (8.58) shows that the sign of the charge produced by the electromagnetic field depends on the edge in question. The charges flowing around the edge of a hole have opposite sign to the charges flowing around the outer edge, and this observation leads to the following exercise.

Exercise 8.6: Consider the behaviour of the edge currents produced by the magnetic field of a solenoid threading the middle of a planar ring-shaped sample (i.e. an annulus). Suppose that the edge currents and the magnetic field of the solenoid are initially zero. Subsequently, the magnetic field is increased from zero to a finite constant, and remains constant thereafter. (a) Show that the total charges of the two edge currents are equal in magnitude but opposite in sign when the magnetic field of the solenoid is in its final state. (b) Explore a thought experiment in which the annulus is deformable. Discuss the behaviour of the edge currents if the annulus is smoothly turned inside-out, i.e. the inner edge becomes the outer edge. Assume that the deformation occurs after the magnetic field has reached its constant value and the magnetic field does not intersect the sample as it is being smoothly deformed.

8.4 BOUNDARY CONDITIONS IN ELECTROMAGNETISM

From a geometrical perspective, the notion of an edge current is a natural consequence of expressing electric current conservation in terms of a distribution. As a result, it should be no surprise that the usual boundary conditions on electric and magnetic fields, typically obtained using "pill-box" arguments in undergraduate physics, naturally emerge when replacing the Maxwell 2-form F with an appropriate 2-distribution F_D. The covariant expressions that readily emerge from this procedure are more than just a succinct way of writing electromagnetic boundary conditions. They readily accommodate boundaries that are accelerating, as well as those that are static (or moving with constant velocity). In particular, they enable us to dispense with the need to Lorentz transform between instantaneous rest frames and the lab frame. For simplicity, we will begin by establishing the main principles of our approach using an integration domain equivalent to a finite subset of $\mathbb{R}^2$. We will then generalise the discussion, using intuitive considerations, to 4-dimensional spacetime.

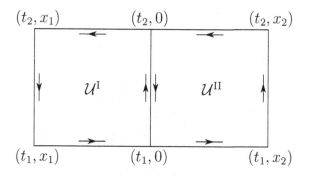

(t_2, x_1) $(t_2, 0)$ (t_2, x_2)

$\mathcal{U}^{\mathrm{I}}$ $\mathcal{U}^{\mathrm{II}}$

(t_1, x_1) $(t_1, 0)$ (t_1, x_2)

Figure 8.3 The region in 2-dimensional spacetime swept out by a pair of connected line segments. Following the usual conventions for spacetime diagrams, time increases in the vertical direction in the diagram. However, again following standard conventions, the time coordinate occupies the first position in the list (t, x), not the second position. Only test forms that vanish on the boundary of the union of $\mathcal{U}^{\mathrm{I}}$ and $\mathcal{U}^{\mathrm{II}}$ are used. The arrows indicate the orientations of the curves $\partial\mathcal{U}^{\mathrm{I}}$ and $\partial\mathcal{U}^{\mathrm{II}}$.

Suppose $\mathcal{U}^{\mathrm{I}}$, $\mathcal{U}^{\mathrm{II}}$ are partitions of a 2-dimensional region in 2-dimensional spacetime. In other words, $\mathcal{U}^{\mathrm{I}}$, $\mathcal{U}^{\mathrm{II}}$ are two 2-dimensional regions whose boundaries $\partial\mathcal{U}^{\mathrm{I}}$, $\partial\mathcal{U}^{\mathrm{II}}$ share a common 1-dimensional component Σ. For example, suppose $\mathcal{U}^{\mathrm{I}}$ is given by $t_1 \leq t \leq t_2$, $x_1 \leq x \leq 0$ and $\mathcal{U}^{\mathrm{II}}$ is given by $t_1 \leq t \leq t_2$, $0 \leq x \leq x_2$. See Figure 8.3. In this case, Σ is the 1-dimensional region (i.e. curve) given by $t_1 \leq t \leq t_2$ at $x = 0$. Let us only consider test forms whose support is entirely inside the union of $\mathcal{U}^{\mathrm{I}}$ and $\mathcal{U}^{\mathrm{II}}$, so their support does not intersect the boundary of the union of $\mathcal{U}^{\mathrm{I}}$ and $\mathcal{U}^{\mathrm{II}}$. In terms of coordinates, the support of the test forms expressed in t, x coordinates is entirely inside the rectangle $t_1 < t < t_2$, $x_1 < x < x_2$.

Based on the above considerations, the integral of a test 1-form over $\partial\mathcal{U}^{\mathrm{I}}$ or $\partial\mathcal{U}^{\mathrm{II}}$ is expressible as an integral of the test 1-form over Σ. However, consideration of the relative orientations of $\partial\mathcal{U}^{\mathrm{I}}$ and $\partial\mathcal{U}^{\mathrm{II}}$ shows that the integrals over Σ must contribute with the same magnitude but opposite sign. In particular, if the boundary $\partial\mathcal{U}_D^{\mathrm{I}}$ of the distribution $\mathcal{U}_D^{\mathrm{I}}$ satisfies $\partial\mathcal{U}_D^{\mathrm{I}} = \Sigma_D$ then it follows $\partial\mathcal{U}_D^{\mathrm{II}} = -\Sigma_D$. From a coordinate-based perspective, $\partial\mathcal{U}^{\mathrm{I}}$ is given by the counter-clockwise perimeter of $t_1 \leq t \leq t_2$, $x_1 \leq x \leq 0$, whereas $\partial\mathcal{U}^{\mathrm{II}}$ is the counter-clockwise perimeter of $t_1 \leq t \leq t_2$, $0 \leq x \leq x_2$. Thus, as anticipated, Σ is given by the straight line from $(t = t_1, x = 0)$ to $(t = t_2, x = 0)$.

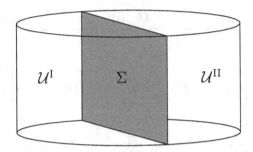

Figure 8.4 A 4-dimensional region in spacetime partitioned by a 3-dimensional region Σ into two 4-dimensional regions $\mathcal{U}^{\mathrm{I}}$ and $\mathcal{U}^{\mathrm{II}}$. We only consider test forms that vanish on the boundary of the union of $\mathcal{U}^{\mathrm{I}}$ and $\mathcal{U}^{\mathrm{II}}$, i.e. the top face, bottom face and curved face of the cylinder in the diagram. However, such test forms are, in general, non-zero on Σ.

The above simple construction is readily generalisable to 4-dimensional spacetime. We only consider test forms whose support is entirely within a particular 4-dimensional region in 4-dimensional spacetime. We partition the 4-dimensional region into two 4-dimensional regions $\mathcal{U}^{\mathrm{I}}$, $\mathcal{U}^{\mathrm{II}}$ by slicing it with a 3-dimensional region Σ, and we choose $\partial \mathcal{U}_D^{\mathrm{I}} = \Sigma_D$. Hence, $\partial \mathcal{U}_D^{\mathrm{II}} = -\Sigma_D$ follows as a consequence. See Figure 8.4.

Maxwell's equations are linear, so we assert the distributional versions

$$dF_D = 0, \qquad d \star F_D = j_D \tag{8.59}$$

of the Gauss-Faraday and Gauss-Ampère laws, respectively (in units where the permittivity and permeability of the vacuum are unity). We will begin by exploring the distributional Gauss-Faraday law, and leave the meaning of the Hodge map of a distribution until it is required during the exploration of the distributional Gauss-Ampère law.

The Maxwell 2-distribution F_D is built from smooth electromagnetic fields in each region; in particular, we introduce

$$F_D = \mathcal{U}_D^{\mathrm{I}} F^{\mathrm{I}} + \mathcal{U}_D^{\mathrm{II}} F^{\mathrm{II}} \tag{8.60}$$

where F^{I} is a Maxwell 2-form in $\mathcal{U}^{\mathrm{I}}$ and F^{II} is a Maxwell 2-form in $\mathcal{U}^{\mathrm{II}}$. However, in addition to receiving contributions from the smooth 3-forms j^{I}, j^{II} in $\mathcal{U}^{\mathrm{I}}$, $\mathcal{U}^{\mathrm{II}}$, respectively, the current 3-distribution j_D includes a term that describes the behaviour of the electric charge and electric current within the interface Σ. It is given as

$$j_D = \mathcal{U}_D^{\mathrm{I}} j^{\mathrm{I}} + \mathcal{U}_D^{\mathrm{II}} j^{\mathrm{II}} + \Sigma_D \wedge j^{\Sigma} \tag{8.61}$$

where, as we will see below, the 2-form j^Σ representing the surface charge and surface current in Σ is determined by the jump in the electromagnetic field across Σ. The distributions $\mathcal{U}_D^I$, $\mathcal{U}_D^{II}$ have degree 0, so we have omitted their wedge products with forms in (8.60) and (8.61).

The exterior derivative of (8.60) is

$$dF_D = -\partial \mathcal{U}_D^I \wedge F^I + \mathcal{U}_D^I dF^I - \partial \mathcal{U}_D^{II} \wedge F^{II} + \mathcal{U}_D^{II} dF^{II} \qquad (8.62)$$

where the substitutions $d\mathcal{U}_D^I = -\partial \mathcal{U}_D^I$ and $d\mathcal{U}_D^{II} = -\partial \mathcal{U}_D^{II}$, which follow because $\mathcal{U}_D^I$ and $\mathcal{U}_D^{II}$ are 0-distributions, have been made. Thus, using $\partial \mathcal{U}_D^I = -\partial \mathcal{U}_D^{II} = \Sigma_D$ gives

$$\Sigma_D \wedge (F^{II} - F^I) + \mathcal{U}_D^I dF^I + \mathcal{U}_D^{II} dF^{II} = 0 \qquad (8.63)$$

since $dF_D = 0$.

The three terms in (8.63) have been presented in a way that emphasises their linear independence. The fact that each term vanishes independently can be deduced by allowing (8.63) to act on certain test 1-forms. In particular, if the test 1-form χ has support fully inside region I then only $(\mathcal{U}_D^I dF^I)[\chi] = 0$ remains from (8.63). This result holds for all χ with support fully inside region I. Likewise, only $(\mathcal{U}_D^I dF^{II})[\chi] = 0$ survives if the support of χ is fully inside region II, and this result holds for all such χ. Moreover, one can choose test forms that go to zero arbitrarily quickly near Σ. Hence, (8.63) yields the Gauss-Faraday law

$$dF^I = 0, \qquad dF^{II} = 0 \qquad (8.64)$$

for the electromagnetic fields inside each region, and the junction condition

$$\Sigma_D \wedge (F^{II} - F^I) = 0 \qquad (8.65)$$

on those fields. Without loss of generality, we can represent the 3-dimensional region Σ as a surface of constant λ, where λ is a scalar field. Thus, $\int_\Sigma d\lambda \wedge \phi = 0$ for all choices of test 2-form ϕ and we conclude that the 1-distribution Σ_D can be expressed as $\Sigma_D = \gamma_D d\lambda$ where γ_D is some 0-distribution. Hence, (8.65) can be written as

$$d\lambda \wedge (F^{II} - F^I) = 0 \quad \text{on } \Sigma \qquad (8.66)$$

or, equivalently,

$$F^{II} - F^I \simeq 0 \qquad (8.67)$$

where $\simeq$ denotes equality after pullback to the interface Σ (i.e. equality up to terms containing $d\lambda$) between regions I and II.

Equation (8.66) (or, equivalently, (8.67)) is a covariant amalgam of the continuity conditions, found in the undergraduate physics textbooks, on the tangential part of the electric field and normal component of the magnetic field. For example, inspection of the expression

$$F = E_x \, dt \wedge dx + E_y \, dt \wedge dy + E_z \, dt \wedge dz$$
$$- B_x \, dy \wedge dz - B_y \, dz \wedge dx - B_z \, dx \wedge dy \qquad (8.68)$$

reveals $F \simeq E_x \, dt \wedge dx + E_y \, dt \wedge dy - B_z \, dx \wedge dy$ on the surface $z = 0$. Hence, (8.67) reveals $E_x^{\mathrm{I}} = E_x^{\mathrm{II}}$, $E_y^{\mathrm{I}} = E_y^{\mathrm{II}}$ and $B_z^{\mathrm{I}} = B_z^{\mathrm{II}}$ at $z = 0$, as anticipated.

The distributional Gauss-Ampère law, $d{\star}F_D = j_D$, yields a covariant amalgam of the standard relationships between surface charge, surface current, electric field and magnetic field that are commonly found using "pill-box" arguments. To proceed further, we need to introduce the action of the Hodge map on distributions. By definition, the Hodge map of a p-distribution T_D is given by

$$\star T_D[\omega] = (-1)^{p(n-p)} T_D[\star\omega], \qquad (8.69)$$

where ω is a test p-form. The following exercise establishes the property of $\star$ required for the subsequent calculation.

Exercise 8.7: Show (8.69) yields the intuitive result $\star(\gamma_D \alpha) = \gamma_D \star \alpha$, where α is any smooth p-form and γ_D is a 0-distribution.

The left-hand side of the distributional Gauss-Ampère law can be quickly unpacked because $\mathcal{U}_D^{\mathrm{I}}$, $\mathcal{U}_D^{\mathrm{II}}$ have degree 0. We have $\star(\mathcal{U}_D^{\mathrm{I}} F^{\mathrm{I}}) = \mathcal{U}_D^{\mathrm{I}} \star F^{\mathrm{I}}$, $\star(\mathcal{U}_D^{\mathrm{II}} F^{\mathrm{II}}) = \mathcal{U}_D^{\mathrm{II}} \star F^{\mathrm{II}}$ and so

$$d \star F_D = -\partial \mathcal{U}_D^{\mathrm{I}} \wedge \star F^{\mathrm{I}} + \mathcal{U}_D^{\mathrm{I}} d \star F^{\mathrm{I}} - \partial \mathcal{U}_D^{\mathrm{II}} \wedge \star F^{\mathrm{II}} + \mathcal{U}_D^{\mathrm{II}} d \star F^{\mathrm{II}} \quad (8.70)$$

follows because $d\mathcal{U}_D^{\mathrm{I}} = -\partial\mathcal{U}_D^{\mathrm{I}}$, $d\mathcal{U}_D^{\mathrm{II}} = -\partial\mathcal{U}_D^{\mathrm{II}}$. Introducing $\partial\mathcal{U}_D^{\mathrm{I}} = \Sigma_D$, $\partial\mathcal{U}_D^{\mathrm{II}} = -\Sigma_D$ to (8.70) and making use of (8.61) yields the result

$$\Sigma_D \wedge \left(\star(F^{\mathrm{II}} - F^{\mathrm{I}}) - j^{\Sigma}\right) + \mathcal{U}_D^{\mathrm{I}}(d \star F^{\mathrm{I}} - j^{\mathrm{I}}) + \mathcal{U}_D^{\mathrm{II}}(d \star F^{\mathrm{II}} - j^{\mathrm{II}}) = 0. \quad (8.71)$$

The argument used to separate the terms in (8.63) is applicable to (8.71). The three terms in (8.71) each vanishes independently, and so we have the Gauss-Ampère law

$$d \star F^{\mathrm{I}} = j^{\mathrm{I}}, \qquad d \star F^{\mathrm{II}} = j^{\mathrm{II}} \qquad (8.72)$$

satisfied by the electromagnetic fields and sources inside each region, and the junction condition

$$\Sigma^D \wedge \star(F^{II} - F^{I}) = \Sigma^D \wedge j^{\Sigma} \tag{8.73}$$

relating the electromagnetic fields at the boundary of each region with the surface charges and surface currents. As we did for the junction condition arising from the distributional Gauss-Faraday law, we can express (8.73) as

$$d\lambda \wedge \star(F^{II} - F^{I}) = d\lambda \wedge j^{\Sigma} \quad \text{on } \Sigma \tag{8.74}$$

or, equivalently,

$$\star(F^{II} - F^{I}) \simeq j^{\Sigma} \tag{8.75}$$

where $\simeq$ denotes equality after pullback to the interface Σ between regions I and II.

Exercise 8.8: Show that (8.74) yields the standard relationships between the jump in the normal component of the electric field, the jump in the tangential components of the magnetic field, the surface charge density and surface current density at the interface $z = 0$ of the regions $z < 0$ and $z > 0$. Convince yourself that (8.75) leads to the same results. You will need to make use of

$$j^{\Sigma} = -\breve{\rho}\, dx \wedge dy + \breve{j}_x\, dt \wedge dy - \breve{j}_y\, dt \wedge dx \tag{8.76}$$

where $\breve{\rho}$ is the surface charge density and $\breve{j}_x$, $\breve{j}_y$ are the Cartesian components of the surface current density at the interface $z = 0$.

8.4.1 Reflection of a pulse from an accelerating mirror

The computational advantage of using (8.67) instead of the non-covariant approach is clear when investigating the reflection of a compact electromagnetic pulse from a perfect, non-stationary, mirror whose velocity is time-dependent. For simplicity, we will suppose that the electromagnetic pulse is a plane wave linearly polarised in the y direction and propagating at normal incidence to the mirror, and the mirror is undergoing arbitrary motion along the x-axis. We saw in a previous chapter that it is fruitful to introduce null coordinates when solving the source-free Maxwell's equations for electromagnetic waves, so we will make use of the null coordinates $x^- = t - x$, $x^+ = t + x$ here. At early times before reflection, the electromagnetic field is given by the 2-form $F_{in} = f_{in}(x^-)dx^- \wedge dy$ of the incoming pulse, where $f_{in}(x^-)$ describes the profile of the incoming pulse. At late times after the pulse is completely reflected, the electromagnetic field is given by the 2-form $F_{out} = f_{out}(x^+)dx^+ \wedge dy$ of the outgoing pulse, where $f_{out}(x^+)$ describes the profile of the outgoing pulse. At intermediate times, the electromagnetic field is given by the superposition of the incoming and outgoing fields, i.e.

$$F = f_{in}(x^-)\,dx^- \wedge dy + f_{out}(x^+)\,dx^+ \wedge dy. \tag{8.77}$$

Clearly, $dF = 0$ follows immediately and $d \star F = 0$ follows almost as quickly by noting $\star(dx^- \wedge dy) = -dx^- \wedge dz$ and $\star(dx^+ \wedge dy) = dx^+ \wedge dz$.

The partition Σ in (8.67) is the 3-dimensional region swept out in spacetime by the surface of the mirror. The mirror is perfectly reflecting, so the pulse cannot penetrate it; hence, $F^{I} = F$ and $F^{II} = 0$, and the junction condition (8.67) reduces to $F \simeq 0$. Although it is possible to introduce a coordinate system adapted to the surface of the mirror, the most efficient way to proceed is to represent the surface of the mirror as an "equipotential" of the 0-form λ given by $\lambda = x - X(t)$. The x coordinate of the surface of the mirror at time t is $X(t)$, so $\lambda = 0$ there. Note that the numerical value of λ on the mirror is unimportant; we only require it to be constant. Hence, $d\lambda \simeq 0$, i.e. $dx \simeq \dot{X}dt$ where the overdot indicates differentiation with respect to t. Note $|\dot{X}| < 1$ because, otherwise, the mirror would be moving at a superluminal speed.

Substituting $dx \simeq \dot{X}dt$ and $x = X(t)$ into (8.77) gives

$$f_{in}(t - X(t))(1 - \dot{X})dt \wedge dy + f_{out}(t + X(t))(1 + \dot{X})dt \wedge dy \simeq 0. \tag{8.78}$$

However, $dt \wedge dy$ is non-zero after pullback to Σ and so

$$f_{out}(s) = -\frac{1 - \dot{X}}{1 + \dot{X}} f_{in}(s - 2X(t))\Big|_{t=t(s)} \tag{8.79}$$

where $t(s)$ is a solution to the implicit equation $s = t + X(t)$. In the simplest case, $X(t) = vt$ and thus

$$f_{out}(s) = -\frac{1 - v}{1 + v} f_{in}\left(\frac{1 - v}{1 + v}s\right). \tag{8.80}$$

Inspection of (8.80) shows that each Fourier mode of the outgoing pulse is a double Doppler-shifted version of a mode of the incoming pulse. To see this, note that the temporal Fourier transform of (8.80) satisfies

$$\int_{-\infty}^{\infty} e^{-i\Omega t} f_{\text{out}}(t)\, dt = -\int_{-\infty}^{\infty} e^{-i\omega t} f_{\text{in}}(t)\, dt \qquad (8.81)$$

where the angular frequency Ω of each outgoing Fourier mode is related to the angular frequency ω of its corresponding incoming mode according to

$$\Omega = \frac{1-v}{1+v}\omega. \qquad (8.82)$$

Since $F^{\text{I}} = F$ and $F^{\text{II}} = 0$, the condition (8.75) satisfied by the surface current 2-form j^{Σ} at the mirror reduces to

$$j^{\Sigma} \simeq - \star F \qquad (8.83)$$

where F is given by (8.77) and f_{out} is determined using (8.79). An explicit expression for j^{Σ} is established in the following exercise.

Exercise 8.9: (a) Show that the 1-form $\mathbf{e} = (\dot{X} dx - dt)/\sqrt{1 - \dot{X}^2}$ is timelike, unit normalised and orthogonal to $d\lambda$, where $\lambda = x - X(t)$.
(b) Show

$$j^{\Sigma} = -2 f_{\text{in}}(t - X(t)) \sqrt{\frac{1 - \dot{X}}{1 + \dot{X}}}\, \mathbf{e} \wedge dz \qquad (8.84)$$

is the unique solution to (8.83) that satisfies $i_Y j^{\Sigma} = 0$, where the vector field $Y = \widetilde{d\lambda}$ is the metric dual of the 1-form $d\lambda$. Discuss the physical meaning of the $\mathbf{e} \wedge dz$ component of j^{Σ}.

Hint: For the very last part of the exercise, note $j^{\Sigma} = -\check{\rho}\, dy \wedge dz + \check{j}_y\, dt \wedge dz - \check{j}_z\, dt \wedge dy$ is the general expression for the surface current 2-form when the mirror is at rest.

Solutions and hints to the exercises

A.1 DIFFERENTIAL FORMS

1.1. The temperature T is constant so $\int_1^2 T\,dS = T(S_2 - S_1)$, and using $P = RT/V$ yields $\int_1^2 P\,dV = RT\ln(V_2/V_1)$. The required result follows immediately.

1.2. Substituting

$$dT = (\partial T/\partial P)_S\,dP + (\partial T/\partial S)_P\,dS, \qquad (A.1)$$

$$dV = (\partial V/\partial P)_S\,dP + (\partial V/\partial S)_P\,dS \qquad (A.2)$$

in $dT \wedge dS = dP \wedge dV$, and using $dS \wedge dS = dP \wedge dP = 0$, gives the first result. The working towards the second result is essentially identical, although the substitutions

$$dS = (\partial S/\partial V)_T\,dV + (\partial S/\partial T)_V\,dT, \qquad (A.3)$$

$$dP = (\partial P/\partial V)_T\,dV + (\partial P/\partial T)_V\,dT \qquad (A.4)$$

are used alongside $dV \wedge dV = dT \wedge dT = 0$. The third result follows in a similar fashion using

$$dS = (\partial S/\partial P)_T\,dP + (\partial S/\partial T)_P\,dT, \qquad (A.5)$$

$$dV = (\partial V/\partial P)_T\,dP + (\partial V/\partial T)_P\,dT \qquad (A.6)$$

and $dP \wedge dP = dT \wedge dT = 0$. The minus sign arises because the 1-forms in the wedge product need to follow the same sequence on both sides of the equation before their coefficients can be equated. This can be arranged by using, for example, $dT \wedge dP = -dP \wedge dT$.

DOI: 10.1201/9781003228943-A

1.3. The first result follows immediately from

$$dT = (\partial T/\partial V)_S\, dV + (\partial T/\partial S)_V\, dS, \tag{A.7}$$
$$dV = (\partial V/\partial T)_S\, dT + (\partial V/\partial S)_T\, dS \tag{A.8}$$

and $dS \wedge dS = 0$. Expanding the brackets and using $dT \wedge dT = dV \wedge dV = 0$ and $dV \wedge dT = -dT \wedge dV$ gives the second result.

1.4. Begin by noting $d(dU) = 0$ yields

$$d((\partial U/\partial S)_V\, dS) = -d((\partial U/\partial V)_S\, dV). \tag{A.9}$$

Then note

$$d((\partial U/\partial S)_V\, dS) = d((\partial U/\partial S)_V) \wedge dS + (\partial U/\partial S)_V d(dS)$$
$$= d((\partial U/\partial S)_V) \wedge dS \tag{A.10}$$

since $d(dS) = 0$. Furthermore,

$$d((\partial U/\partial S)_V) = \frac{\partial(\partial U/\partial S)}{\partial V}\, dV + \left(\frac{\partial^2 U}{\partial S^2}\right)_V dS \tag{A.11}$$

and so

$$d((\partial U/\partial S)_V\, dS) = \frac{\partial(\partial U/\partial S)}{\partial V}\, dV \wedge dS \tag{A.12}$$

because $dS \wedge dS = 0$. Likewise

$$d((\partial U/\partial V)_S\, dV) = \frac{\partial(\partial U/\partial V)}{\partial S}\, dS \wedge dV \tag{A.13}$$

Thus, plugging (A.12) and (A.13) into (A.9), and using $dS \wedge dV = -dV \wedge dS$, reveals

$$\frac{\partial(\partial U/\partial S)}{\partial V} = \frac{\partial(\partial U/\partial V)}{\partial S}. \tag{A.14}$$

A.2 VECTOR FIELDS AND THEIR RELATIONSHIP WITH DIFFERENTIAL FORMS

2.1. The expression $X_V V = \zeta\,(\partial V/\partial V)_P + \xi\,(\partial V/\partial P)_V$ and the condition $X_V V = 0$ yield $\zeta = 0$. Thus $X_V T = \xi\,(\partial T/\partial P)_V$, and the condition $X_V T = 1$ gives $\xi = 1/(\partial T/\partial P)_V$ as required. The expression for X_P is determined in a similar fashion.

2.2. The result $dV(X_V) = X_V V$ follows immediately from the definition of the contraction of dV and X_V. Thus, $dV(X_V) = 0$ is a consequence of the condition $X_V V = 0$. Likewise, $dP(X_V) = X_V P$ and using the explicit expression for X_V reveals $dP(X_V) = 1/(\partial T/\partial P)_V$. The results of the contractions involving X_P follow in a similar manner.

2.3. Since X_P satisfies the conditions $X_P T = 1$ and $X_P P = 0$, we immediately obtain $X_P = \partial/\partial T$ in the $P - T$ plane. The explicit expression for X_V requires more algebra, but it is straightforward to obtain. Using $X_V = \check{\zeta}\partial/\partial T + \check{\xi}\partial/\partial P$ (where carons have been introduced to avoid a conflict with the notation used in a previous solution) and the condition $X_V T = 1$, we find $\check{\zeta} = 1$ because $(\partial T/\partial P)_T = 0$. The condition $X_V V = 0$ can then be expressed as $(\partial V/\partial T)_P + \check{\xi}(\partial V/\partial P)_T = 0$, giving the required result for $\check{\xi}$. The relationship $1/(\partial T/\partial V)_P = (\partial V/\partial T)_P$ follows immediately by comparing the results $dV(X_P) = 1/(\partial T/\partial V)_P$ and $X_P V = (\partial V/\partial T)_P$. The remainder of the question follows in the same manner, except the working is undertaken in the $T - V$ plane instead of the $P - T$ plane.

A.3 ASPECTS OF INTEGRATION

3.1. Begin by noting

$$d(k_x\, dy \wedge dz) = dk_x \wedge dy \wedge dz + k_x\, d(dy) \wedge dz - k_x\, dy \wedge d(dz)$$
$$= dk_x \wedge dy \wedge dz \qquad \text{(A.15)}$$

where the first equality follows using the graded Leibniz rule and the second equality holds because $d(dy) = 0$ and $d(dz) = 0$. Furthermore, $dk_x = \partial_x k_x dx + \partial_y k_x dy + \partial_x k_x dz$ and, hence,

$$d(k_x\, dy \wedge dz) = \partial_x k_x\, dx \wedge dy \wedge dz + \partial_y k_x\, dy \wedge dy \wedge dz$$
$$+ \partial_z k_x\, dz \wedge dy \wedge dz$$
$$= \partial_x k_x\, dx \wedge dy \wedge dz \qquad \text{(A.16)}$$

where the second equality holds because $dy \wedge dy = 0$ and furthermore $dz \wedge dy \wedge dz = -dz \wedge dz \wedge dy = 0$. Likewise,

$$d(k_y\, dz \wedge dx) = \partial_y k_y\, dy \wedge dz \wedge dx, \qquad \text{(A.17)}$$
$$d(k_z\, dx \wedge dy) = \partial_z k_z\, dz \wedge dx \wedge dy. \qquad \text{(A.18)}$$

The required result follows by inspecting dk and noting

$$dy \wedge dz \wedge dx = dx \wedge dy \wedge dz = dz \wedge dx \wedge dy \qquad (A.19)$$

follows from the antisymmetry of the wedge product of two 1-forms.

3.2. The first result can be obtained by introducing the representation $\theta \mapsto (x = r\cos\theta, y = r\sin\theta)$, $0 \le \theta \le 2\pi$, of a circle C of radius r whose centre is at the origin. It follows

$$\int_C x\, dy = \int_0^{2\pi} r\cos\theta\, d(r\sin\theta)$$
$$= \int_0^{2\pi} r^2 \cos^2\theta\, d\theta$$
$$= \pi r^2 \qquad (A.20)$$

which, as stated in the question, is the area of C. The second result, $\int_C y\, dx = -\pi r^2$, follows in the same fashion.

3.3. Replacing y with $y(u,v)$ and z with $z(u,v)$ gives

$$\int_\Sigma j_x\, dy \wedge dz = \int_D j_x(\partial_u y\, du + \partial_v y\, dv) \wedge (\partial_u z\, du + \partial_v z\, dv) \qquad (A.21)$$

and so

$$\int_\Sigma j_x\, dy \wedge dz = \int_D j_x(\partial_u y\, \partial_v z - \partial_u z\, \partial_v y)\, du \wedge dv \qquad (A.22)$$

where $du \wedge du = 0$, $dv \wedge dv = 0$ and $du \wedge dv = -dv \wedge du$ have been used after expanding the brackets. It follows

$$\int_\Sigma j_x\, dy \wedge dz = \int_D j_x\, \mathbf{e}_x \cdot (\partial_u \mathbf{r} \times \partial_v \mathbf{r})\, du \wedge dv \qquad (A.23)$$

and, likewise,

$$\int_\Sigma j_y\, dz \wedge dx = \int_D j_y\, \mathbf{e}_y \cdot (\partial_u \mathbf{r} \times \partial_v \mathbf{r})\, du \wedge dv, \qquad (A.24)$$
$$\int_\Sigma j_z\, dx \wedge dy = \int_D j_z\, \mathbf{e}_z \cdot (\partial_u \mathbf{r} \times \partial_v \mathbf{r})\, du \wedge dv. \qquad (A.25)$$

Combining the above results, and deleting the wedge from $du \wedge dv$, yields the outcome stated in the question.

3.4. One way to proceed is to begin by noting

$$d(A_x\, dx) = dA_x \wedge dx + A_x\, d(dx)$$
$$= (\partial_x A_x dx + \partial_y A_x dy + \partial_z A_x dz) \wedge dx$$
$$= \partial_y A_x\, dy \wedge dx + \partial_z A_x\, dz \wedge dx$$

where $d(dx) = 0$ and $dx \wedge dx = 0$ have been used. Likewise, it follows

$$d(A_y dy) = \partial_x A_y\, dx \wedge dy + \partial_z A_y\, dz \wedge dy, \qquad (A.26)$$
$$d(A_z dz) = \partial_x A_z\, dx \wedge dz + \partial_y A_z\, dy \wedge dz \qquad (A.27)$$

and, using $dy \wedge dx = -dx \wedge dy$ and $dz \wedge dy = -dy \wedge dz$, yields

$$d\alpha = (\nabla \times \mathbf{A})_x\, dy \wedge dz + (\nabla \times \mathbf{A})_y\, dz \wedge dx + (\nabla \times \mathbf{A})_z\, dx \wedge dy \quad (A.28)$$

where $(\nabla \times \mathbf{A})_x = \mathbf{e}_x \cdot (\nabla \times \mathbf{A})$, $(\nabla \times \mathbf{A})_y = \mathbf{e}_y \cdot (\nabla \times \mathbf{A})$ and $(\nabla \times \mathbf{A})_z = \mathbf{e}_z \cdot (\nabla \times \mathbf{A})$. Thus, substituting $k = d\alpha$ (i.e. $\mathbf{k} = \nabla \times \mathbf{A}$) in Exercise 3.1 reveals the required result.

A.4 THE METRIC TENSOR

4.1. Applying the rank $(0,2)$ tensor g to the pair ∂_a, ∂_b of vector fields gives

$$g(\partial_a, \partial_b) = (g_{cd}\, dx^c \otimes dx^d)(\partial_a, \partial_b)$$
$$= g_{cd}\, dx^c(\partial_a)\, dx^d(\partial_b)$$
$$= g_{cd}\, \delta^c_a\, \delta^d_b$$
$$= g_{ab} \qquad (A.29)$$

as required.

4.2. The metric g is positive definite, and so any new coframe $\{\check{e}^1, \check{e}^2\}$ related to the orthonormal coframe $\{e^1, e^2\}$ by a rotation will be orthonormal. Introducing

$$\check{e}^1 = \cos\theta\, e^1 + \sin\theta\, e^2, \qquad \check{e}^2 = -\sin\theta\, e^1 + \cos\theta\, e^2, \qquad (A.30)$$

where θ is a 0-form, gives

$$\check{e}^1 \otimes \check{e}^1 = (\cos\theta\, e^1 + \sin\theta\, e^2) \otimes (\cos\theta\, e^1 + \sin\theta\, e^2)$$
$$= \cos^2\theta\, e^1 \otimes e^1 + \sin^2\theta\, e^2 \otimes e^2 \qquad (A.31)$$
$$+ \sin\theta \cos\theta\, (e^1 \otimes e^2 + e^2 \otimes e^1)$$

and

$$\check{e}^2 \otimes \check{e}^2 = (-\sin\theta\, e^1 + \cos\theta\, e^2) \otimes (-\sin\theta\, e^1 + \cos\theta\, e^2)$$
$$= \sin^2\theta\, e^1 \otimes e^1 + \cos^2\theta\, e^2 \otimes e^2 \qquad \text{(A.32)}$$
$$- \sin\theta\cos\theta\,(e^1 \otimes e^2 + e^2 \otimes e^1).$$

Thus, the metric $g = e^1 \otimes e^1 + e^2 \otimes e^2$ can also be expressed as

$$g = \check{e}^1 \otimes \check{e}^1 + \check{e}^2 \otimes \check{e}^2, \qquad \text{(A.33)}$$

i.e. as asserted earlier, $\{\check{e}^1, \check{e}^2\}$ is orthonormal because $\{e^1, e^2\}$ is orthonormal.

The frame $\{X_1, X_2\}$ and coframe $\{e^1, e^2\}$ are dual, and thus satisfy $e^a(X_b) = \delta^a_b$. In other words,

$$e^1(X_1) = 1, \quad e^2(X_1) = 0, \quad e^1(X_2) = 0, \quad e^2(X_2) = 1. \quad \text{(A.34)}$$

In principle, given any coframe in any number of dimensions, one can use matrix methods to calculate its dual frame. However, in the present case, it is instructive to use a more primitive approach. Inspection of the expression for e^2 shows X_1 must satisfy $X_1 = \varepsilon\,\partial_x$ for some function ε of (x, y); otherwise, $e^2(X_1) = 0$ will not hold. The requirement $e^1(X_1) = 1$ fixes ε, giving the result

$$X_1 = \frac{1}{\alpha}\frac{\partial}{\partial x}. \qquad \text{(A.35)}$$

The vector field X_2 must be a linear combination of ∂_x and ∂_y with non-zero coefficients; otherwise, $e^1(X_2) = 0$ will not hold. The contraction $e^2(X_2) = 1$ is satisfied by

$$X_2 = \zeta\frac{\partial}{\partial x} + (\gamma^2 - (\beta/\alpha)^2)^{-1/2}\frac{\partial}{\partial y} \qquad \text{(A.36)}$$

for some function ζ of (x, y), which is fixed using $e^1(X_2) = 0$. The result is

$$\zeta = -\frac{\beta}{\alpha^2}(\gamma^2 - (\beta/\alpha)^2)^{-1/2}. \qquad \text{(A.37)}$$

Clearly, both components of X_2 diverge if $\beta = \alpha\gamma$. Note that e^2 vanishes in this case, so the pair $\{e^1, e^2\}$ does not span the space of covectors, i.e. it is not a coframe.

Another perspective on this problem emerges by noting that the vector field

$$V = -\frac{\beta}{\alpha}\frac{\partial}{\partial x} + \alpha\frac{\partial}{\partial y} \tag{A.38}$$

satisfies $g(V, W) = 0$ *for all* vector fields W when $\beta = \alpha\gamma$. Thus, the $(0, 2)$-tensor g does not satisfy the basic properties of a metric if $\beta = \alpha\gamma$; in particular, it is degenerate. The notion of *orthonormality* is a metric-based concept, and so it is reasonable that no orthonormal coframe (or orthonormal frame) exists when $\beta = \alpha\gamma$.

4.3. We have the metric $g = \eta_{ab}\, \bar{e}^a \otimes \bar{e}^b$; hence $g = \eta_{ab}\, \mathcal{O}^a{}_c\mathcal{O}^b{}_d e^c \otimes e^d$. Inspection of $g = \eta_{cd}\, e^c \otimes e^d$ then yields the required result.

4.4. Inspection of the orthonormal coframe

$$\left\{ e^1 = \alpha dx + \frac{\beta}{\alpha}dy, \quad e^2 = \sqrt{\left(\frac{\beta}{\alpha}\right)^2 - \gamma^2 dy} \right\} \tag{A.39}$$

shows $g = e^1 \otimes e^1 - e^2 \otimes e^2$. Hence, the signature of g is 0.

4.5. Since

$$\gamma_1' = \frac{d\gamma_1^a}{d\lambda}\frac{\partial}{\partial x^a} \tag{A.40}$$

and, using the chain rule,

$$\frac{d\gamma_1^a}{d\lambda} = \frac{d\rho}{d\lambda}\frac{d\gamma_1^a}{d\rho} \tag{A.41}$$

we have

$$\gamma_1' = \frac{d\rho}{d\lambda}\dot{\gamma}_1 \tag{A.42}$$

where

$$\dot{\gamma}_1 = \frac{d\gamma_1^a}{d\rho}\frac{\partial}{\partial x^a}. \tag{A.43}$$

However, tensor contraction is a linear process, so $g(\gamma_1', \gamma_2') = fg(\dot{\gamma}_1, \gamma_2')$ where $f = d\rho/d\lambda$. Likewise, $g(\gamma_1', \gamma_1') = f^2 g(\dot{\gamma}_1, \dot{\gamma}_1)$. Inspection of the expression for the angle θ shows that it is independent of f; thus, the required result is obtained.

4.6. Begin by noting $\{X_0 = \partial/\partial t, X_1 = \partial/\partial x\}$ is an orthonormal frame; thus $g(\gamma_1', \gamma_1') = g(X_0, X_0) = -1$ and

$$g(\gamma_2', \gamma_2') = \frac{1}{1 - v^2} \Big(g(X_0, X_0) + vg(X_0, X_1)$$

$$+ vg(X_1, X_0) + v^2 g(X_1, X_1) \Big)$$

$$= -1. \tag{A.44}$$

Likewise, $g(\gamma_1', \gamma_2') = -1/\sqrt{1 - v^2}$ and we obtain the expression $w = \cosh^{-1}(1/\sqrt{1 - v^2})$ for the rapidity.

4.7. Suppose γ_1 and γ_2 intersect at the point p. The tangent vector γ_1' is timelike, and an orthonormal frame $\{X_0, X_1, \ldots, X_{n-1}\}$ exists such that

$$X_0 = \frac{\gamma_1'}{\sqrt{-g(\gamma_1', \gamma_1')}} \tag{A.45}$$

at p. As stated in the question, the tangent vectors γ_1' and γ_2' are orthogonal at p; thus, it follows $g(X_0, \gamma_2') = 0$ at p. Hence, when expressed as a linear combination of the frame elements, the vector γ_2' cannot contain X_0; otherwise, $g(X_0, \gamma_2')$ would be non-zero at p because $g(X_0, X_0) \neq 0$. The vector γ_2' must be a linear combination of the remaining $n-1$ frame elements, i.e. $\{X_1, \ldots, X_n\}$. However, all of those remaining elements are spacelike because, as stated in the question, the spacetime is Lorentzian. The vector X_0 is the sole timelike element of the orthonormal frame. Any linear combination of $\{X_1, \ldots, X_{n-1}\}$, such as γ_2', is spacelike.

4.8. Both coframes $\{f^1, f^2\}$ and $\{e^1, e^2\}$ are orthonormal, so they must be related by a rotation and/or a reflection. If the transformation is a pure rotation then

$$f^1 = \cos\theta\, e^1 + \sin\theta\, e^2, \quad f^2 = -\sin\theta\, e^1 + \cos\theta\, e^2 \tag{A.46}$$

where θ is any 0-form. It follows

$$f^1 \wedge f^2 = (\cos\theta\, e^1 + \sin\theta\, e^2) \wedge (-\sin\theta\, e^1 + \cos\theta\, e^2)$$

$$= -\sin\theta\cos\theta\, e^1 \wedge e^1 + \cos^2\theta\, e^1 \wedge e^2$$

$$- \sin^2\theta\, e^2 \wedge e^1 + \sin\theta\cos\theta\, e^2 \wedge e^2$$

$$= e^1 \wedge e^2 \tag{A.47}$$

where $e^1 \wedge e^1 = e^2 \wedge e^2 = 0$ and $e^2 \wedge e^1 = -e^1 \wedge e^2$ have been used in the final step. An essentially identical calculation yields $f^1 \wedge f^2 = -e^1 \wedge e^2$ if the coframes are related by a rotation and a reflection, i.e.

$$f^1 = -\sin\theta \, e^1 + \cos\theta \, e^2, \quad f^2 = \cos\theta \, e^1 + \sin\theta \, e^2. \qquad (A.48)$$

4.9. Although the calculation can be undertaken using Cartesian coordinates, it is more convenient to use spherical polar coordinates. We have

$$x = r\sin\theta\cos\varphi, \quad y = r\sin\theta\sin\varphi, \quad z = r\cos\theta \qquad (A.49)$$

where $0 < \theta < \pi$ and $0 < \varphi < 2\pi$. The induced metric is given by setting $\Phi = 0$, i.e. $r = a$, and $d\Phi = 0$, i.e. $dr = 0$, in the Euclidean metric $g_{\text{Euclid}} = dr \otimes dr + r^2 \, d\theta \otimes d\theta + r^2 \sin^2\theta \, d\varphi \otimes d\varphi$. Hence, the induced metric on the 2-sphere S^2 is

$$g_{S^2} = a^2 \, d\theta \otimes d\theta + a^2 \sin^2\theta \, d\varphi \otimes d\varphi \qquad (A.50)$$

and, by inspection, $\{e^1 = a \, d\theta, \; e^2 = a\sin\theta \, d\varphi\}$ is an orthonormal coframe. Thus, the volume (i.e. area) of the 2-sphere is $\int_{S^2} e^1 \wedge e^2 = \int_{S^2} a^2 \sin\theta \, d\theta \wedge d\varphi$. However,

$$\int_{S^2} \sin\theta \, d\theta \wedge d\varphi = \int_0^{2\pi} \int_0^{\pi} \sin\theta \, d\theta \, d\varphi = 4\pi \qquad (A.51)$$

and so the required "volume" is $4\pi a^2$.

4.10. The exterior derivative of $p_0 = \sqrt{m^2 + \delta_{ij} \, u^i u^j}$ is

$$dp_0 = \frac{1}{2} \frac{\delta_{ij} du^i u^j + \delta_{ij} u^i du^j}{\sqrt{m^2 + \delta_{kl} \, u^k u^l}} \qquad (A.52)$$

where the indices on the terms inside the square root have been relabelled to avoid a clash with the indices on the terms in the numerator. Swapping the repeated indices on $\delta_{ij} u^i du^j$ gives $\delta_{ji} u^j du^i$; however, the Kronecker delta satisfies $\delta_{ji} = \delta_{ij}$. Hence,

$$dp_0 = \frac{u_i du^i}{\sqrt{m^2 + \delta_{kl} \, u^k u^l}} \qquad (A.53)$$

where $u_i = \delta_{ij} u^j$. Thus, substituting dp_0 into the metric $\hat{g}$ given in Eq. (4.45) and collecting terms yields the required result for g.

4.11. Since $\widetilde{V}(Y) = g(V, Y)$ for any vector field Y, we have

$$\widetilde{V}(Y) = g_{ab}\, e^a(V)\, e^b(Y). \qquad (A.54)$$

The above must hold for all Y, and so the first required result

$$\widetilde{V} = g_{ab}\, V^a\, e^b \qquad (A.55)$$

is obtained, where $V^a = e^a(V)$ has been used.

Likewise, $\widetilde{\alpha}$ satisfies $\alpha(Y) = g(\widetilde{\alpha}, Y)$ for any vector field Y. Hence

$$\alpha(Y) = g_{ab}\, e^a(\widetilde{\alpha})\, e^b(Y) \qquad (A.56)$$

and thus, using $\alpha = \alpha_b\, e^b$, we obtain $\alpha_b = g_{ab}\, e^a(\widetilde{\alpha})$ by inspection. We can isolate $e^a(\widetilde{\alpha})$ by making use of the components g^{cb} of the inverse metric tensor. Since $g^{cb} g_{ba} = \delta^c_a$ we find $g^{cb}\alpha_b = e^c(\widetilde{\alpha})$ and the second required result

$$\widetilde{\alpha} = g^{cb}\alpha_b\, X_c \qquad (A.57)$$

is obtained. Hence, setting $\alpha = \widetilde{V}$ and noting $\widetilde{V}_b = g_{ab}V^a$ from (A.55) immediately yields the third required result.

4.12. The required result can be obtained using inductive reasoning. Begin by assuming α is decomposable, i.e. it can be written as the wedge product of 1-forms. To be precise, $\alpha = \beta_1 \wedge \beta_2 \wedge \cdots \wedge \beta_p$ where $\{\beta_1, \beta_2, \ldots, \beta_p\}$ is a set of p 1-forms. Inspection of the rule $\star(\alpha \wedge \beta) = i_{\widetilde{\beta}} \star \alpha$ shows that

$$\deg\big(\star(\alpha \wedge \beta)\big) = \deg(\star\alpha) - 1 \qquad (A.58)$$

where $\deg(\gamma)$ denotes the degree of the differential form γ. Thus, if $\deg(\star\alpha) = n - p$, where α is a p-form, then (A.58) yields $\deg(\star(\alpha \wedge \beta)) = (n - p) - 1$. In other words, if $\star\alpha$ is a $(n - p)$-form then $\star(\alpha \wedge \beta)$ is a $(n - q)$-form where $q = p + 1$. However, if α is a 0-form then (A.58) immediately leads to $\deg(\star\beta) = \deg(\star 1) - 1$ because the degree of a differential form is inert under multiplication by a 0-form. Furthermore, the volume form $\star 1$ has degree n and so $\deg(\star\beta) = n - 1$. Hence, $\deg(\star\alpha) = n - \deg(\alpha)$ must hold.

The above argument can be immediately extended to non-decomposable p-forms, i.e. a p-form whose explicit expression can only be written as the sum of at least two decomposable p-forms.

Every differential p-form, where $n > p > 1$, can be expressed as a linear combination of wedge products of p coframe elements. For example, every 2-form in 3-dimensional Euclidean space can be expressed as a linear combination of $dx \wedge dy$, $dy \wedge dz$ and $dz \wedge dx$. The degree of any sum of p-forms is p, and we see that the required result holds for all differential forms.

4.13. By definition, the quantity $G_2(\alpha \wedge \beta, \gamma \wedge \delta)$ is determined by

$$\alpha \wedge \beta \wedge \star(\gamma \wedge \delta) = G_2(\alpha \wedge \beta, \gamma \wedge \delta) \star 1. \qquad \text{(A.59)}$$

The left-hand side of the equality can be unravelled by beginning with $\beta \wedge \star(\gamma \wedge \delta)$. We have

$$\begin{aligned}
\beta \wedge \star(\gamma \wedge \delta) &= \beta \wedge i_{\tilde{\delta}} \star \gamma \\
&= -i_{\tilde{\delta}}(\beta \wedge \star \gamma) + i_{\tilde{\delta}}\beta \star \gamma \\
&= -G_1(\beta, \gamma) i_{\tilde{\delta}} \star 1 + G_1(\beta, \delta) \star \gamma \\
&= -G_1(\beta, \gamma) \star \delta + G_1(\beta, \delta) \star \gamma. \qquad \text{(A.60)}
\end{aligned}$$

Hence

$$\begin{aligned}
\alpha \wedge \beta \wedge \star(\gamma \wedge \delta) &= \alpha \wedge \left(-G_1(\beta, \gamma) \star \delta + G_1(\beta, \delta) \star \gamma \right) \\
&= -G_1(\beta, \gamma) G_1(\alpha, \delta) \star 1 + G_1(\alpha, \gamma) G_1(\beta, \delta) \star 1 \\
&\qquad\qquad\qquad\qquad\qquad\qquad\qquad\qquad\qquad \text{(A.61)}
\end{aligned}$$

and we arrive at the result

$$G_2(\alpha \wedge \beta, \gamma \wedge \delta) = G_1(\alpha, \gamma) G_1(\beta, \delta) - G_1(\beta, \gamma) G_1(\alpha, \delta). \quad \text{(A.62)}$$

The inverse metric tensor satisfies $G_1(\alpha, \beta) = G_1(\beta, \alpha)$, and so the identity $G_2(\alpha \wedge \beta, \gamma \wedge \delta) = G_2(\gamma \wedge \delta, \alpha \wedge \beta)$ is revealed.

4.14. Inspection of the metric reveals the orthonormal coframe

$$\left\{ e^1 = \frac{m}{\sqrt{m^2 + \rho^2}} d\rho, e^2 = \rho \, d\theta, e^3 = \rho \sin \theta \, d\varphi \right\} \qquad \text{(A.63)}$$

and so the electric density 2-form can be written as

$$\begin{aligned}
j = \frac{j_\rho}{\rho^2 \sin \theta} e^2 \wedge e^3 &+ \frac{j_\theta \sqrt{m^2 + \rho^2}}{m\rho \sin \theta} e^3 \wedge e^1 \\
&+ \frac{j_\varphi \sqrt{m^2 + \rho^2}}{m\rho} e^1 \wedge e^2. \qquad \text{(A.64)}
\end{aligned}$$

Let $\{X_1, X_2, X_3\}$ be the orthonormal frame dual to $\{e^1, e^2, e^3\}$. Thus, $\widetilde{e^1} = X_1$ and the introduction of the volume 3-form $\star 1 = e^1 \wedge e^2 \wedge e^3$ reveals $\star e^1 = i_{X_1} \star 1 = e^2 \wedge e^3$. Likewise, $\widetilde{e^2} = X_2$ and it follows $\star(e^1 \wedge e^2) = i_{X_2} \star e^1 = e^3$. The results $\star(e^2 \wedge e^3) = e^1$ and $\star(e^3 \wedge e^1) = e^2$ follow in a similar manner since $\widetilde{e^3} = X_3$. Hence

$$\mathcal{J} = \frac{j_\rho}{\rho^2 \sin\theta} X_1 + \frac{j_\theta \sqrt{m^2 + \rho^2}}{m\rho \sin\theta} X_2 + \frac{j_\varphi \sqrt{m^2 + \rho^2}}{m\rho} X_3. \quad (A.65)$$

Introduction of the explicit expressions

$$\left\{ X_1 = \frac{\sqrt{m^2 + \rho^2}}{m} \frac{\partial}{\partial\rho}, X_2 = \frac{1}{\rho} \frac{\partial}{\partial\theta}, X_3 = \frac{1}{\rho \sin\theta} \frac{\partial}{\partial\varphi} \right\} \quad (A.66)$$

allows $\mathcal{J}$ to be expressed as

$$\mathcal{J} = \frac{\sqrt{m^2 + \rho^2}}{m\rho^2 \sin\theta} \left(j_\rho \frac{\partial}{\partial\rho} + j_\theta \frac{\partial}{\partial\theta} + j_\varphi \frac{\partial}{\partial\varphi} \right). \quad (A.67)$$

4.15. Contracting the relationship $g' = \Omega^2 g$ on the pair (X_a, X_b) yields $g'(X_a, X_b) = \Omega^2 g(X_a, X_b)$, i.e. $g'_{ab} = \Omega^2 g_{ab}$ where $g' = g'_{ab} e^a \otimes e^b$ and $g = g_{ab} e^a \otimes e^b$. However, $g'^{ab} g'_{bc} = \delta^a_c$ and thus $\Omega^2 g'^{ab} g_{bc} = \delta^a_c$. By inspection of this result, it is clear that $g^{ab} = \Omega^2 g'^{ab}$, i.e. $G_1 = \Omega^2 G'_1$ as required.

Since $G_2(e^a \wedge e^b, e^c \wedge e^d) = g^{ac} g^{bd} - g^{bc} g^{ad}$, we see $G'_2(e^a \wedge e^b, e^c \wedge e^d) = \Omega^{-4}(g^{ac} g^{bd} - g^{bc} g^{ad})$, i.e. $G'_2 = \Omega^{-4} G_2$. Likewise, the inner products G'_3 and G_3 on 3-forms satisfy $G'_3 = \Omega^{-6} G_3$. This pattern continues to higher degree forms, i.e. the inner products G'_p and G_p on p-forms satisfy $G'_p = \Omega^{-2p} G_p$.

Although this result can be proven explicitly, it is instructive to deduce it based on intuitive reasoning. Begin by noting that the quantity $G_p(\alpha_1 \wedge \cdots \wedge \alpha_p, \beta_1 \wedge \cdots \wedge \beta_p)$ is a 0-form. However, using the properties of the Hodge map, it must be possible to express this quantity entirely in terms of G_1 and the 1-forms $\alpha_1, \ldots, \alpha_p, \beta_1, \ldots, \beta_p$.

Every 1-form in the above list must appear once in every term in the final unravelled result, because contraction is a linear operation and the final result is a 0-form. Thus, the inner product on 1-forms must appear p times in each term in the final unravelled result. Therefore, a comparison of the final unravelled expressions for the contractions of G'_p and G_p on the above list of 1-forms must lead to $G'_p = (\Omega^{-2})^p G_p$.

4.16. Introducing $dt = a(t)d\tau$ in g_{FLRW} with $k = 0$ gives

$$
\begin{aligned}
g_{\text{FLRW}} = a(t)^2(&-d\tau \otimes d\tau + dr \otimes dr \\
&+ r^2 d\theta \otimes d\theta + r^2 \sin^2\theta \, d\varphi \otimes d\varphi).
\end{aligned} \tag{A.68}
$$

Thus, $g_{\text{FLRW}} = \Omega^2 g_{\text{Mink}}$ where $\Omega = a(t)$ and

$$
\begin{aligned}
g_{\text{Mink}} &= -d\tau \otimes d\tau + dr \otimes dr \\
&\quad + r^2 d\theta \otimes d\theta + r^2 \sin^2\theta \, d\varphi \otimes d\varphi \\
&= -d\tau \otimes d\tau + dx \otimes dx + dy \otimes dy + dz \otimes dz
\end{aligned} \tag{A.69}
$$

with

$$
x = r\sin\theta\cos\varphi, \qquad y = r\sin\theta\sin\varphi, \qquad z = r\cos\theta. \tag{A.70}
$$

The *conformal time* τ satisfies $\tau = \int dt/a(t)$. For example, for a matter-dominated universe $\tau = 3t^{1/3}/A$ whilst for a dark-energy-dominated universe $\tau = -e^{-Ht}/C$ where $C = BH$.

4.17. The Helmholtz equation in 2 dimensions can be written as $d \star d\phi + k^2\phi \star 1 = 0$. Use of $\star'\alpha = \Omega^{(2-2p)} \star \alpha$ shows that the two terms in the Helmholtz equation do not scale in the same manner. In particular, although $\star'd\phi = \star d\phi$, we find $\star'1 = \Omega^2 \star 1$. Thus, the Helmholtz equation in 2 dimensions is not conformally invariant.

4.18. Rule 3 satisfied by the Lie derivative leads to

$$
\begin{aligned}
\mathcal{L}_{\partial/\partial T} g = \mathcal{L}_{\partial/\partial T} \big(&- (1 + \alpha X)^2 \big) \, dT \otimes dT \\
&- (1 + \alpha X)^2 \left(\mathcal{L}_{\partial/\partial T} \, dT \otimes dT + dT \otimes \mathcal{L}_{\partial/\partial T} \, dT \right) \\
&+ \mathcal{L}_{\partial/\partial T} \, dX \otimes dX + dX \otimes \mathcal{L}_{\partial/\partial T} \, dX.
\end{aligned} \tag{A.71}
$$

The first term in the above expression is zero using rule 1 and noting $\partial X/\partial T = 0$. The remaining terms vanish using rule 2, followed by rule 1 and the fact that $\partial X/\partial T$ and $\partial T/\partial T$ are both constant. Thus, as asserted in the question, $\partial/\partial T$ is a Killing vector.

The expression

$$
\begin{aligned}
\frac{\partial}{\partial T} &= \frac{\partial t}{\partial T} \frac{\partial}{\partial t} + \frac{\partial x}{\partial T} \frac{\partial}{\partial x} \\
&= \alpha \left(x \frac{\partial}{\partial t} + t \frac{\partial}{\partial x} \right)
\end{aligned} \tag{A.72}
$$

follows from Eq. (4.54) using the chain rule for partial derivatives. We see that $\partial/\partial T$ is a constant multiple of the boost Killing vector K.

A.5 MAXWELL'S EQUATIONS IN TERMS OF DIFFERENTIAL FORMS

5.1. We have $\widetilde{\mathcal{H}} = H_x dx + H_y dy + H_z dz$ and so

$$d\widetilde{\mathcal{H}} = dH_x \wedge dx + dH_y \wedge dy + dH_z \wedge dz \qquad (A.73)$$

where, for example, $d(H_x dx) = dH_x \wedge dx + H_x d(dx)$ and $d(dx) = 0$ have been used. Using $dH_x = \partial_x H_x dx + \partial_y H_x dy + \partial_z H_x dz$ gives

$$dH_x \wedge dx = \frac{\partial H_x}{\partial y} dy \wedge dx + \frac{\partial H_x}{\partial z} dz \wedge dx \qquad (A.74)$$

where $dx \wedge dx = 0$ has been used. The same approach also yields

$$dH_y \wedge dy = \frac{\partial H_y}{\partial x} dx \wedge dy + \frac{\partial H_y}{\partial z} dz \wedge dy, \qquad (A.75)$$

$$dH_z \wedge dz = \frac{\partial H_z}{\partial x} dx \wedge dz + \frac{\partial H_z}{\partial y} dy \wedge dz \qquad (A.76)$$

and it follows

$$d\widetilde{\mathcal{H}} = \left(\frac{\partial H_y}{\partial x} - \frac{\partial H_x}{\partial y}\right) dx \wedge dy + \left(\frac{\partial H_z}{\partial y} - \frac{\partial H_y}{\partial z}\right) dy \wedge dz$$
$$+ \left(\frac{\partial H_x}{\partial z} - \frac{\partial H_z}{\partial x}\right) dz \wedge dx \qquad (A.77)$$

where $dy \wedge dx = -dx \wedge dy$, $dx \wedge dz = -dz \wedge dx$ and $dz \wedge dy = -dy \wedge dz$ have been used. However, $\star dx = dy \wedge dz$, $\star dy = dz \wedge dx$ and $\star dz = dx \wedge dy$, and so

$$d\widetilde{\mathcal{H}} = j_z \star dz + j_x \star dx + j_y \star dy \qquad (A.78)$$

where $\nabla \times \mathbf{H} = \mathbf{j}$ has been used. The right-hand side of (A.78) is $\star\widetilde{\mathcal{J}}$.

5.2. Since $d(t\,dx) = dt \wedge dx + t\,d(dx) = dt \wedge dx$, we have $d(dt \wedge dx) = d(d(t\,dx)) = 0$. Thus, $d(E_x dt \wedge dx) = dE_x \wedge dt \wedge dx + E_x d(dt \wedge dx) = dE_x \wedge dt \wedge dx$. Furthermore, $dE_x = \partial_t E_x dt + \partial_x E_x dx + \partial_y E_x dy + \partial_z E_x dz$ and, since $dt \wedge dt = dx \wedge dx = 0$, we have $dE_x \wedge dt \wedge dx =$

$(\partial_y E_x dy + \partial_z E_x dz) \wedge dt \wedge dx$. Expanding the brackets and using $dy \wedge dt \wedge dx = dt \wedge dx \wedge dy$ and $dz \wedge dt = -dt \wedge dz$ yields the first required result. The second required result is obtained using a similar method.

Applying the cyclic permutation $x \mapsto y$, $y \mapsto z$, $z \mapsto x$ to the first required result gives

$$d(E_y \, dt \wedge dy) = \partial_z E_y \, dt \wedge dy \wedge dz - \partial_x E_y \, dt \wedge dx \wedge dy, \quad \text{(A.79)}$$
$$d(E_z \, dt \wedge dz) = \partial_x E_z \, dt \wedge dz \wedge dx - \partial_y E_z \, dt \wedge dy \wedge dz \quad \text{(A.80)}$$

and it follows

$$d(E_x \, dt \wedge dx + E_y \, dt \wedge dy + E_z \, dt \wedge dz) = -(\boldsymbol{\nabla} \times \mathbf{E})_z \, dt \wedge dx \wedge dy$$
$$-(\boldsymbol{\nabla} \times \mathbf{E})_x \, dt \wedge dy \wedge dz - (\boldsymbol{\nabla} \times \mathbf{E})_y \, dt \wedge dz \wedge dx.$$

Likewise, combining the results obtained from applying the cyclic permutation $x \mapsto y$, $y \mapsto z$, $z \mapsto x$ to the second required result gives

$$d(B_x \, dy \wedge dz + B_y \, dz \wedge dx + B_z \, dx \wedge dy) = (\partial_t \mathbf{B})_x \, dt \wedge dy \wedge dz$$
$$+(\partial_t \mathbf{B})_y \, dt \wedge dz \wedge dx + (\partial_t \mathbf{B})_z \, dt \wedge dx \wedge dy$$
$$+\boldsymbol{\nabla} \cdot \mathbf{B} \, dx \wedge dy \wedge dz.$$

Thus, $dF = 0$ yields Faraday's law of induction and Gauss's law for magnetism.

5.3. The key subtlety is to appreciate $\widetilde{dt} = -\partial_t$, whereas $\widetilde{dx} = \partial_x$, $\widetilde{dy} = \partial_y$ and $\widetilde{dz} = \partial_z$. The required results then emerge by application of the rules satisfied by the interior product. For example, $\star dt = i_{\widetilde{dt}} \star 1 = -i_{\partial_t} \star 1 = -dx \wedge dy \wedge dz$ since $\star 1 = dt \wedge dx \wedge dy \wedge dz$. Hence, $\star(dt \wedge dx) = i_{\widetilde{dx}} \star dt = i_{\partial_x} \star dt$ and so $\star(dt \wedge dx) = -dy \wedge dz$ as asserted in the question. The remaining results are obtained in a similar fashion.

Inspection of the results given in the question shows that the operator $- \star \star$ acts as the identity on all of them. Any 2-form can be expressed as a linear superposition of the 2-form basis elements given in the question, and so the final required result is obtained. Substituting $\det\eta = -1$, $p = 2$ and $n = 4$ in Eq. (4.67) yields the same result.

5.4. The first required result follows immediately from the linearity of the Hodge map and use of the stated results. The second required result can be quickly obtained by comparing the components of $\star F$ and F, and inspecting the results found in previous exercises. The substitutions $\mathbf{E} \mapsto -\mathbf{B}$, $\mathbf{B} \mapsto \mathbf{E}$ are equivalent to the substitution $F \mapsto \star F$, and making this substitution in $dF = 0$ immediately yields the second required result.

5.5. Inspection of the expressions for F and $\star F$ show that the terms in $F \wedge \star F$ combine in pairs to give a result proportional to $dt \wedge dx \wedge dy \wedge dz$. For example, the term $E_x dt \wedge dx$ in F combines with the term $-E_x dy \wedge dz$ in $\star F$. All other possible terms containing E_x vanish because the wedge product of a 1-form with itself vanishes. Similar results hold for the remaining five non-zero pairs of terms. It follows $F \wedge \star F = (|\mathbf{B}|^2 - |\mathbf{E}|^2) \, dt \wedge dx \wedge dy \wedge dz$. Explicit calculation shows $\star(dt \wedge dx \wedge dy \wedge dz) = -1$ and so $\star(F \wedge \star F) = |\mathbf{E}|^2 - |\mathbf{B}|^2$ as required. A similar calculation reveals $\star(F \wedge F) = 2\mathbf{E} \cdot \mathbf{B}$.

5.6. Comparison of $F = f(u) \, du \wedge dy$ with the general expression for F in terms of the Cartesian components of $\mathbf{E}$ and $\mathbf{B}$ gives the first required result. The second part of the question is a straightforward exercise in vector calculus.

5.7. The first required result follows immediately from the definitions of P and $\bar{P}$. As noted in the main text, $i\star$ squares to the identity operator on 2-forms. Thus, $(1 - i\star)^2 \beta = 2(1 - i\star)\beta$ where β is a 2-form, and hence $PP = P$. Likewise $(1 - i\star)(1 + i\star)\beta = 0$ follows immediately, i.e. $P\bar{P} = 0$. The complex conjugation of these results immediately yields the remaining two required results.

5.8. The result $d\zeta/\zeta = dr/r + id\theta$ is obtained using the substitution $\zeta = re^{i\theta}$ and, thus, $\mathrm{Re}(\mathfrak{f}(u, \zeta) \, d\zeta) = f(u) \, dr/r$. The required result follows immediately from the expression for F given in the main text.

5.9. Inspection of Eq. (5.15), i.e. the general expression for F in terms of the Cartesian components of $\mathbf{E}$ and $\mathbf{B}$, gives $i_{\partial_t} F = E_x dx + E_y dy + E_z dz$. Tensor contraction is a linear process, and so the metric dual of $i_{\partial_t} F$ satisfies $\widetilde{i_{\partial_t} F} = E_x \widetilde{dx} + E_y \widetilde{dy} + E_z \widetilde{dz}$. However, we have $\widetilde{dx} = \partial_x$, $\widetilde{dy} = \partial_y$ and $\widetilde{dz} = \partial_z$ and the expression for $\mathcal{E}$ follows from the definition of $\mathcal{E}$ given in the main text. The expression for

$\mathcal{B}$ follows in a similar manner using the general expression for $\star F$ in terms of the Cartesian components of $\mathbf{E}$ and $\mathbf{B}$.

5.10. The required result can be obtained by combining the results found in previous exercises.

5.11. The radial coordinate is constant, and so

$$dx = r_0 \cos \theta \cos \varphi \, d\theta - r_0 \sin \theta \sin \varphi \, d\varphi, \tag{A.81}$$
$$dy = r_0 \cos \theta \sin \varphi \, d\theta + r_0 \sin \theta \cos \varphi \, d\varphi, \tag{A.82}$$
$$dz = -r_0 \sin \theta \, d\theta. \tag{A.83}$$

The first three required results follow by making use of $d\theta \wedge d\theta = d\varphi \wedge d\varphi = 0$ and $d\varphi \wedge d\theta = -d\theta \wedge d\varphi$. It is straightforward to see that the final required result holds if $r = r_0$. Furthermore, the additional terms indicated by ... must contain dr because they do not appear when r is constant.

5.12. The required result follows by integrating the continuity equation over the ball and the specified time interval. The divergence theorem is used to express the result in terms of an integral over the surface of the ball. This exercise does not make use of differential forms, so the details are left to the reader.

5.13. Begin by noting

$$i_{\partial_t} \langle F \rangle = \langle E_x \rangle dx + \langle E_y \rangle dy + \langle E_z \rangle dz, \tag{A.84}$$
$$\langle F \rangle - dt \wedge i_{\partial_t} \langle F \rangle = -\langle B_x \rangle dy \wedge dz - \langle B_y \rangle dz \wedge dx$$
$$- \langle B_z \rangle dx \wedge dy. \tag{A.85}$$

Thus,

$$G = dt \wedge (\epsilon_r \, i_{\partial_t} \langle F \rangle) + \frac{1}{\mu_r} (\langle F \rangle - dt \wedge i_{\partial_t} \langle F \rangle) \tag{A.86}$$

and the required result follows immediately.

A.6 CLASSICAL MECHANICS

6.1. The scalar lift of the 2-form $\beta = \beta_{ab} \, dx^a \wedge dx^b$ is $\beta^* = \beta_{ab} v^a v^b$. However, due to the antisymmetry of the wedge product, the components of β can be chosen to satisfy $\beta_{ab} = -\beta_{ba}$ without loss of generality. Hence, $\beta^* = -\beta_{ba}v^a v^b = -\beta_{ab}v^b v^a$ where the repeated indices have been relabelled in the second step. Since $v^b v^a = v^a v^b$ we arrive at $\beta^* = -\beta^*$, i.e. $\beta^* = 0$.

Inspection of $g = g_{ab}\,dx^a \otimes dx^b$ shows its scalar lift is $g^* = g_{ab}v^a v^b$.

6.2. Substituting $v_\theta = a/r^2$ in V and noting $v_r = \dot{r}$ reveals $\dot{V} = 0$. The required expression for v_θ follows immediately from rearranging the second item in Eq. (6.14). The required expression for v_r then follows by substituting v_θ in Eq. (6.15) and solving for v_r. Note there are two possible solutions for v_r (the ambiguity raised in the question).

6.3. The required result follows immediately from differentiating $\frac{1}{2}mv^2 + V$ with respect to t and making use of Eq. (6.19) while noting $v = \dot{x}$.

6.4. Any 1-form $\alpha = \alpha_a dq^a$ on M can be understood as the section $\{q^a, p_b = \alpha_b(q^1, \ldots, q^n)\}$ of the cotangent bundle T^*M. The 1-form α can also be expressed as $\{q'^a, p'_b = \alpha'_b(q'^1, \ldots, q'^n)\}$ where $\{\alpha'_b\}$ are the components of α in another coordinate system $\{q'^a\}$ on M. However, $\alpha_a = \alpha'_b \partial_a q'^b$ where $\partial_a q'^b$ denotes the partial derivative of the bth primed coordinate on M with respect to the ath unprimed coordinate on M. Thus, we see $p_a = p'_b \partial_a q'^b$ must hold and, hence, $\theta = p_a dq^a = p'_a dq'^a$ as required.

Likewise, any vector field V on M can be understood as the section $\{x^a, v^b = V^b(x^1, \ldots, x^n)\}$ of the tangent bundle TM. The vector field V can also be expressed as $\{x'^a, v'^b = V'^b(x'^1, \ldots, x'^n)\}$, and thus the relationship $v'^a = v^b \partial_b x'^a$ must hold, where $\partial_b x'^a$ denotes the partial derivative of the ath primed coordinate on M with respect to the bth unprimed coordinate on M. It is clear that, in general, $\sum_{a=1}^n v^a dx^a \neq \sum_{a=1}^n v'^a dx'^a$.

6.5. Using $\omega = dq^a \wedge dp_a$ and introducing

$$\alpha^\sharp = A^a \frac{\partial}{\partial q^a} + B_a \frac{\partial}{\partial p_a} \qquad (A.87)$$

gives

$$i_{\alpha^\sharp}\omega = i_{\alpha^\sharp}dq^a\,dp_a - i_{\alpha^\sharp}dp_a\,dq^a$$
$$= A^a dp_a - B_a dq^a. \qquad (A.88)$$

Inspection of $\alpha = i_{\alpha^\sharp}\omega$ reveals $A^a = \gamma^a$ and $B_a = -\beta_a$.

6.6. The canonical momentum is

$$p_a = \frac{\partial L}{\partial v^a} = mv_a + qA_a \qquad (A.89)$$

where $v_a = \delta_{ab} v^b$, since $g_{ab} = \delta_{ab}$ in Cartesian coordinates in 3-dimensional Euclidean space. Expressed using standard vector calculus notation, the canonical momentum is $\mathbf{p} = m\mathbf{v} + e\mathbf{A}$ where $m\mathbf{v}$ is the kinetic momentum.

6.7. Explicit calculation shows

$$
\begin{aligned}
i_{X_H}\omega &= i_{X_H} dq^a \, dp_a - i_{X_H} dp_a \, dq^a \\
&= X_H q^a \, dp_a - X_H p_a \, dq^a \\
&= \frac{\partial H}{\partial p_a} dp_a + \frac{\partial H}{\partial q^a} dq^a
\end{aligned}
\qquad (A.90)
$$

as required.

6.8. Substituting the given Hamiltonian into Eq. (6.43) yields

$$X_H = \frac{p}{m}\frac{\partial}{\partial q} - kq\frac{\partial}{\partial p}. \qquad (A.91)$$

It immediately follows $X_H H = 0$, i.e. $\mathcal{L}_{X_H} H = 0$ using rule 1 for the Lie derivative (see Chapter 4). A fundamental property of the Lie derivative is that it commutes with the exterior derivative (rule 2), and so $\mathcal{L}_{X_H} dq = d(\mathcal{L}_{X_H} q) = d(X_H q) = dp/m$. Likewise, $\mathcal{L}_{X_H} dp = -k\, dq$. Hence, it follows $\mathcal{L}_{X_H}(dq \wedge dp) = \mathcal{L}_{X_H} dq \wedge dp + dq \wedge \mathcal{L}_{X_H} dp = 0$ because $dq \wedge dq = dp \wedge dp = 0$. Thus, $\mathcal{L}_{X_H}\omega = 0$. Furthermore, the cotangent bundle T^*M is 2-dimensional, and so $\Omega = \omega$. The result $\mathcal{L}_{X_H}\Omega = 0$ follows immediately.

6.9. Begin by noting Eq. (6.54) gives

$$
\begin{aligned}
\{f, \{g, h\}\} &= \partial_C f \, \mathcal{I}^{CD} \partial_D (\partial_A g \, \mathcal{I}^{AB} \partial_B h) \\
&= \partial_C f \, \mathcal{I}^{CD} \partial_D \partial_A g \, \mathcal{I}^{AB} \partial_B h \\
&\quad + \partial_C f \, \mathcal{I}^{CD} \partial_A g \, \partial_D \mathcal{I}^{AB} \partial_B h \\
&\quad + \partial_C f \, \mathcal{I}^{CD} \partial_A g \, \mathcal{I}^{AB} \partial_D \partial_B h
\end{aligned}
\qquad (A.92)
$$

where ∂_A is shorthand for $\partial/\partial z^A$. The final term in (A.92) can be written as

$$
\begin{aligned}
\partial_C f \, \mathcal{I}^{CD} \partial_A g \, \mathcal{I}^{AB} \partial_D \partial_B h &= \partial_B f \, \mathcal{I}^{BA} \partial_C g \, \mathcal{I}^{CD} \partial_A \partial_D h \\
&= -\partial_C g \, \mathcal{I}^{CD} \partial_D \partial_A h \, \mathcal{I}^{AB} \partial_B f
\end{aligned}
\qquad (A.93)
$$

using the substitutions $C \mapsto B$, $D \mapsto A$, $A \mapsto C$, $B \mapsto D$ in the first step and using the identities $\mathcal{I}^{BA} = -\mathcal{I}^{AB}$ and $\partial_A \partial_D h = \partial_D \partial_A h$ in the second step. We see that the third term in the expansion of $\{f, \{g, h\}\}$ is the negative of the first term in the expansion of $\{g, \{h, f\}\}$. Consideration of this result under the cyclic permutation $f \mapsto g \mapsto h \mapsto f$ shows that six of the nine terms in the expansion of the Poisson brackets in the Jacobi identity cancel in pairs. The remaining three terms are

$$
\begin{aligned}
0 = {} & \partial_C f \, \mathcal{I}^{CD} \partial_A g \, \partial_D \mathcal{I}^{AB} \partial_B h \\
& + \partial_C g \, \mathcal{I}^{CD} \partial_A h \, \partial_D \mathcal{I}^{AB} \partial_B f \\
& + \partial_C h \, \mathcal{I}^{CD} \partial_A f \, \partial_D \mathcal{I}^{AB} \partial_B g \quad \text{(A.94)}
\end{aligned}
$$

which, after some index relabelling, can be concisely expressed as

$$
(\mathcal{I}^{AD} \partial_D \mathcal{I}^{BC} + \mathcal{I}^{BD} \partial_D \mathcal{I}^{CA} + \mathcal{I}^{CD} \partial_D \mathcal{I}^{AB}) \, \partial_A f \, \partial_B g \, \partial_C h = 0. \quad \text{(A.95)}
$$

6.10. Let f be a function of (x, y, z) only. The explicit expression

$$
\{f, g\} = \frac{\partial f}{\partial q^a} \frac{\partial g}{\partial p_a} - \frac{\partial f}{\partial p_a} \frac{\partial g}{\partial q^a} \quad \text{(A.96)}
$$

for the Poisson bracket $\{f, g\}$ reduces to

$$
\{f, g\} = \frac{\partial f}{\partial x} \frac{\partial g}{\partial p_x} + \frac{\partial f}{\partial y} \frac{\partial g}{\partial p_y} + \frac{\partial f}{\partial z} \frac{\partial g}{\partial p_z} \quad \text{(A.97)}
$$

because $\partial f / \partial p_x = \partial f / \partial p_y = \partial f / \partial p_z = 0$. Introducing $g = x p_y - y p_x$ gives

$$
\{f, g\} = x \frac{\partial f}{\partial y} - y \frac{\partial f}{\partial x}. \quad \text{(A.98)}
$$

However, the change of variable $x = r \cos\theta$, $y = r \sin\theta$ yields $\partial f / \partial \theta = x \, \partial f / \partial y - y \, \partial f / \partial x$ and we see $\{f, g\}$ is zero if f does not depend on θ. We conclude g generates rotations around the z-axis.

6.11. It is straightforward to see that the explicit expression

$$
\{f, g\} = \frac{\partial f}{\partial x} \frac{\partial g}{\partial p_x} + \frac{\partial f}{\partial y} \frac{\partial g}{\partial p_y} - \frac{\partial f}{\partial p_x} \frac{\partial g}{\partial x} - \frac{\partial f}{\partial p_y} \frac{\partial g}{\partial y} \quad \text{(A.99)}
$$

for the Poisson bracket of any pair of 0-forms f and g yields $\{p_x, x p_y\} = -p_y$ and $\{p_x, y p_x\} = 0$. Furthermore, it is clear

$\{p_y, H\} = 0$ because the Hamiltonian H given in the question does not depend on y. Hence, $\{\{f, g\}, H\} = 0$ for the choices of f and g given in the question. This result can be expressed as $X_H\{f, g\} = 0$, i.e. $d\{f, g\}/dt = 0$.

6.12. The original Lagrangian is immediately obtained because $dt/d\lambda = 1$ in this case.

6.13. The action is the integral of the 1-form $-m\sqrt{-\eta_{ab}\dot{x}^a\dot{x}^b}dt$. However, $\sqrt{-\eta_{ab}\dot{x}^a\dot{x}^b}dt = \sqrt{-\eta_{ab}x'^a x'^b}d\lambda$ because $\dot{x}^a = x'^a\dot{\lambda}$ and $d\lambda = \dot{\lambda}dt$. We see that the Lagrangian is already invariant under reparameterisations.

The canonical momentum is

$$p_a = \frac{\partial \tilde{L}}{\partial \dot{x}^a} \tag{A.100}$$

$$= \frac{m\eta_{ab}\dot{x}^b}{\sqrt{-\eta_{cd}\dot{x}^c\dot{x}^d}} \tag{A.101}$$

and so $p_a\dot{x}^a = -m\sqrt{-\eta_{ab}\dot{x}^a\dot{x}^b}$. It follows that the natural Hamiltonian $\tilde{H} = p_a\dot{x}^a - \tilde{L}$ vanishes.

Inspection of the expression for p_a shows $\eta^{ab}p_ap_b = -m^2$. Hence, we introduce the relation $\phi = \eta^{ab}p_ap_b + m^2$ because $\phi \approx 0$ on the space of physical solutions. Use of the total Hamiltonian $\tilde{H}_T = v\phi$ yields

$$\dot{x}^a = \{x^a, \tilde{H}_T\} \approx 2v\eta^{ab}p_b, \quad \dot{p}_a = \{p_a, \tilde{H}_T\} \approx 0. \tag{A.102}$$

The choice $2v = 1/m$ corresponds to proper time parameterisation, whereas the choice $2v = 1/p^0$ corresponds to lab time parameterisation (where $p^0 = \eta^{0b}p_b$).

6.14. We will determine the details of the Poisson bracket using the second suggested method, i.e. use the identity $\{f, g\} = dg^{\sharp}f$, where $dg = i_{dg^{\sharp}}\omega$. Begin by noting

$$i_{dg^{\sharp}}\omega = i_{dg^{\sharp}}dy^a \, d\pi_a - i_{dg^{\sharp}}d\pi_a \, dy^a - qB_{ba} \, i_{dg^{\sharp}}dy^b \, dy^a \tag{A.103}$$

using the expansion $B = \frac{1}{2}B_{ba} \, dy^b \wedge dy^a$ of the magnetic 2-form B and the rules of the interior product. Furthermore,

$$dg = \frac{\partial g}{\partial y^a} \, dy^a + \frac{\partial g}{\partial \pi_a} \, d\pi_a \tag{A.104}$$

and thus, using $dg = i_{dg^\sharp}\omega$, we find

$$\frac{\partial g}{\partial y^a} = -i_{dg^\sharp}d\pi_a - qB_{ba}\,i_{dg^\sharp}dy^b, \quad \frac{\partial g}{\partial \pi_a} = i_{dg^\sharp}dy^a. \qquad (A.105)$$

Therefore, recalling the identity $dh(X) = i_X dh$ where h is any 0-form and X is any vector field, we obtain

$$dg^\sharp = i_{dg^\sharp}dy^a \frac{\partial}{\partial y^a} + i_{dg^\sharp}d\pi_a \frac{\partial}{\partial \pi_a}$$

$$= \frac{\partial g}{\partial \pi_a}\frac{\partial}{\partial y^a} - \left(\frac{\partial g}{\partial y^a} + qB_{ba}\frac{\partial g}{\partial \pi_b}\right)\frac{\partial}{\partial \pi_a} \qquad (A.106)$$

and, since $\{f, g\} = dg^\sharp f$, it follows

$$\{f, g\} = \frac{\partial f}{\partial y^a}\frac{\partial g}{\partial \pi_a} - \frac{\partial f}{\partial \pi_a}\frac{\partial g}{\partial y^a} + qB_{ab}\frac{\partial f}{\partial \pi_a}\frac{\partial g}{\partial \pi_b} \qquad (A.107)$$

where $B_{ba} = -B_{ab}$ has been used in the final step.

6.15. An efficient way to obtain the required result is to make use of the general expression

$$\mathcal{J} = \{f, \{g, h\}\} + \{g, \{h, f\}\} + \{h, \{f, g\}\} \qquad (A.108)$$

for the Jacobiator found in the working in Exercise 6.9. As shown previously, the introduction of

$$\{f, g\} = \frac{\partial f}{\partial z^A}\mathcal{I}^{AB}\frac{\partial g}{\partial z^B} \qquad (A.109)$$

yields

$$\mathcal{J} = (\mathcal{I}^{AD}\partial_D\mathcal{I}^{BC} + \mathcal{I}^{BD}\partial_D\mathcal{I}^{CA} + \mathcal{I}^{CD}\partial_D\mathcal{I}^{AB})\partial_A f\,\partial_B g\,\partial_C h. \qquad (A.110)$$

Let $\{\mathcal{I}^{ab}\}$, $\{\mathcal{I}^a{}_b\}$, $\{\mathcal{I}_a{}^b\}$, $\{\mathcal{I}_{ab}\}$ be the elements of $\{\mathcal{I}^{AB}\}$ corresponding to the four different pairs of coordinates arising from (y^a, π_b), i.e. y-y, y-π, π-y and π-π. Thus, (A.109) can be written as

$$\{f, g\} = \frac{\partial f}{\partial y^a}\mathcal{I}^{ab}\frac{\partial g}{\partial y^b} + \frac{\partial f}{\partial y^a}\mathcal{I}^a{}_b\frac{\partial g}{\partial \pi_b}$$

$$+ \frac{\partial f}{\partial \pi_a}\mathcal{I}_a{}^b\frac{\partial g}{\partial y^b} + \frac{\partial f}{\partial \pi_a}\mathcal{I}_{ab}\frac{\partial g}{\partial \pi_b}, \qquad (A.111)$$

where $\mathcal{I}^{ab} = 0$, $\mathcal{I}^a{}_b = \delta^a_b$, $\mathcal{I}_a{}^b = -\delta^b_a$ and $\mathcal{I}_{ab} = qB_{ab}$ follow from inspection of (A.107). The first term

$$\mathcal{I}^{AD}\partial_D\mathcal{I}^{BC}\partial_A f\,\partial_B g\,\partial_C h = -q\frac{\partial B_{bc}}{\partial y^a}\frac{\partial f}{\partial \pi_a}\frac{\partial g}{\partial \pi_b}\frac{\partial h}{\partial \pi_c} \qquad (A.112)$$

in the expression (A.110) for $\mathcal{J}$ then readily emerges. The remaining two terms in $\mathcal{J}$ can be obtained using the cyclic permutation $f \mapsto g \mapsto h \mapsto f$. The final result can be expressed as

$$\mathcal{J} = -q\left(\frac{\partial B_{bc}}{\partial y^a} + \frac{\partial B_{ca}}{\partial y^b} + \frac{\partial B_{ab}}{\partial y^c}\right)\frac{\partial f}{\partial \pi_a}\frac{\partial g}{\partial \pi_b}\frac{\partial h}{\partial \pi_c}. \qquad (A.113)$$

6.16. The exterior derivative of the symplectic 2-form ω is

$$d\omega = \lambda dp \wedge dq \wedge dt. \qquad (A.114)$$

By definition, the Hamiltonian vector field $dH^\sharp$ is determined by $dH = i_{dH^\sharp}\omega$. Unpacking both sides of $dH = i_{dH^\sharp}\omega$ using the rules of exterior calculus gives

$$dp_t + \frac{p}{m}dp + kq\,dq = i_{dH^\sharp}dq\,dp - i_{dH^\sharp}dp\,dq$$
$$+ i_{dH^\sharp}dt\,dp_t - i_{dH^\sharp}dp_t\,dt$$
$$+ \lambda p\,i_{dH^\sharp}dq\,dt - \lambda p\,i_{dH^\sharp}dt\,dq. \qquad (A.115)$$

Equating the coefficients of dq, dt, dp, dp_t yields

$$kq = -i_{dH^\sharp}dp - \lambda p\,i_{dH^\sharp}dt, \quad 0 = -i_{dH^\sharp}dp_t + \lambda p\,i_{dH^\sharp}dq, \quad (A.116)$$
$$\frac{p}{m} = i_{dH^\sharp}dq, \quad 1 = i_{dH^\sharp}dt. \qquad (A.117)$$

Hence

$$i_{dH^\sharp}dq = \frac{p}{m}, \quad i_{dH^\sharp}dt = 1, \qquad (A.118)$$

$$i_{dH^\sharp}dp = -(\lambda p + kq), \quad i_{dH^\sharp}dp_t = \frac{\lambda p^2}{m}, \qquad (A.119)$$

and so the Hamiltonian vector field is

$$dH^\sharp = i_{dH^\sharp}dq\frac{\partial}{\partial q} + i_{dH^\sharp}dt\frac{\partial}{\partial t} + i_{dH^\sharp}dp\frac{\partial}{\partial p} + i_{dH^\sharp}dp_t\frac{\partial}{\partial p_t}$$

$$= \frac{p}{m}\frac{\partial}{\partial q} + \frac{\partial}{\partial t} - (\lambda p + kq)\frac{\partial}{\partial p} + \frac{\lambda p^2}{m}\frac{\partial}{\partial p_t}. \qquad (A.120)$$

Using Cartan's identity, $\mathcal{L}_{dH\sharp}\omega = di_{dH\sharp}\omega + i_{dH\sharp}d\omega = i_{dH\sharp}d\omega$ because $di_{dH\sharp}\omega = d(dH) = 0$. Thus, (A.114) and (A.120) give

$$\mathcal{L}_{dH\sharp}\omega = -\lambda(\lambda p + kq)\, dq \wedge dt - \frac{\lambda p}{m}\, dp \wedge dt + \lambda\, dp \wedge dq. \quad \text{(A.121)}$$

The cotangent bundle T^*M is 4-dimensional, and so the volume form is $\Omega = \omega \wedge \omega$; hence, $\mathcal{L}_{dH\sharp}\Omega = 2\omega \wedge \mathcal{L}_{dH\sharp}\omega$. On expanding the explicit result, inspection of the wedge products shows that almost all of the terms vanish except

$$\mathcal{L}_{dH\sharp}\Omega = 2\lambda\, dt \wedge dp_t \wedge dp \wedge dq. \quad \text{(A.122)}$$

However, $\Omega = 2\, dq \wedge dp \wedge dt \wedge dp_t$ and we arrive at the result

$$\mathcal{L}_{dH\sharp}\Omega = -\lambda\Omega. \quad \text{(A.123)}$$

We conclude that the phase space volume occupied by a collection of solutions to Hamilton's equations exponentially decays at rate λ along the solutions. In fact, it is straightforward to see that λ is a decay constant by inspecting Hamilton's equations. For example, we have

$$\dot{q} = \{q, H\} = dH^{\sharp}q = \frac{p}{m} \quad \text{(A.124)}$$

and, likewise,

$$\dot{p} = -\lambda p - kq, \quad \dot{t} = 1, \quad \dot{p}_t = \frac{\lambda p^2}{m} \quad \text{(A.125)}$$

follows from the remaining components of $dH^{\sharp}$. Clearly, the oscillator's equation of motion is $m\ddot{q} + m\lambda\dot{q} + kq = 0$.

A.7 CONNECTIONS

7.1. The Lie derivative of a vector field is antisymmetric in its arguments, i.e. $\mathcal{L}_X Y = -\mathcal{L}_Y X$. Thus $\mathcal{L}_V V$ is identically zero, regardless of the details of V.

Furthermore, the acceleration of a fluid in Newtonian physics is given by the expression $(\mathbf{v} \cdot \boldsymbol{\nabla})\mathbf{v}$ when the flow velocity $\mathbf{v}$ is steady. Clearly $(f\mathbf{v} \cdot \boldsymbol{\nabla})\mathbf{v} = f(\mathbf{v} \cdot \boldsymbol{\nabla})\mathbf{v}$, whereas $\mathcal{L}_{fV} V \neq f\mathcal{L}_V V$ in general.

7.2. Since $\nabla_X X_a = \omega^b{}_a(X) X_b$ we have $e^b(\nabla_X X_a) = \omega^b{}_a(X)$. However, $e^b(\nabla_X X_a) = \nabla_X(e^b(X_a)) - (\nabla_X e^b)(X_a)$ because ∇_X commutes with tensor contractions. The first term vanishes because the 0-forms $\{e^a(X_b)\}$ are constant. Introducing $\nabla_X e^b = \Omega^b{}_c(X)e^c$ reveals $\omega^b{}_a(X) = -\Omega^b{}_a(X)$, i.e. $\Omega^a{}_b = -\omega^a{}_b$.

7.3. The vector $\dot\gamma$ tangent to the curve $\gamma : \lambda \mapsto x^a = \gamma^a(\lambda)$ is

$$\dot\gamma = \frac{d\gamma^a}{d\lambda}\frac{\partial}{\partial x^a}. \tag{A.126}$$

However, the chain rule

$$\frac{d\gamma^a}{d\lambda} = \frac{d\tilde\lambda}{d\lambda}\frac{d\gamma^a}{d\tilde\lambda} \tag{A.127}$$

leads to the relationship

$$\gamma' = \frac{d\lambda}{d\tilde\lambda}\dot\gamma \tag{A.128}$$

where

$$\gamma' = \frac{d\gamma^a}{d\tilde\lambda}\frac{\partial}{\partial x^a}. \tag{A.129}$$

However, $\nabla_{fX}Y = f\nabla_X Y$ and so $\nabla_{\gamma'}V = 0$ if and only if $\nabla_{\dot\gamma}V = 0$.

7.4. Inspection of $X'_a = \Lambda^b{}_a X_b$ reveals

$$\Lambda^1{}_1 = e^{-F}, \qquad \Lambda^2{}_2 = \frac{e^{-F}}{r}, \qquad \Lambda^1{}_2 = 0, \qquad \Lambda^2{}_1 = 0 \tag{A.130}$$

and so Eq. (7.10) reduces to

$$\omega'^1{}_1 = \omega^1{}_1 - dF, \qquad \omega'^2{}_2 = \omega^2{}_2 - dF - \frac{dr}{r},$$

$$\omega'^1{}_2 = \frac{\omega^1{}_2}{r}, \qquad \omega'^2{}_1 = r\omega^2{}_1. \tag{A.131}$$

Noting $dF = f\,dr$ and substituting the explicit expressions for $\omega^1{}_1$ and $\omega^2{}_2$ yields $\omega'^1{}_1 = 0$ and $\omega'^2{}_2 = 0$. Introducing $V = V^{a'}X'_a$ into Eq. (7.12) and following the same approach used to derive Eq. (7.13) leads to

$$\frac{dV^{1'}}{d\theta} - (1+rf)V^{2'} = 0, \qquad \frac{dV^{2'}}{d\theta} + (1+rf)V^{1'} = 0. \tag{A.132}$$

Thus, we find

$$\frac{d^2 V^{a\prime}}{d\theta^2} + (1 + rf)^2 V^{a\prime} = 0 \qquad (A.133)$$

and arrive at the same conclusion found using the components of V in the frame $\{X_a\}$.

7.5. Tensor contraction is a linear process, and so

$$g(\dot{\gamma}, \dot{\gamma}) = \left(\frac{d\tilde{\lambda}}{d\lambda}\right)^2 g(\gamma', \gamma') \qquad (A.134)$$

where $\dot{\gamma}$ and γ' are related as in the solution to Exercise 7.3. Thus,

$$\int_\alpha^\beta \sqrt{g(\dot{\gamma}, \dot{\gamma})} d\lambda = \int_{\tilde{\alpha}}^{\tilde{\beta}} \sqrt{g(\gamma', \gamma')} d\tilde{\lambda} \qquad (A.135)$$

where $\tilde{\lambda} = \tilde{\alpha}$ at $\lambda = \alpha$ and $\tilde{\lambda} = \tilde{\beta}$ at $\lambda = \beta$.

7.6. If γ is a spacelike curve then the analysis is unchanged. However, if γ is timelike (i.e. $g(\dot{\gamma}, \dot{\gamma}) < 0$) then the appropriate integral to render stationary is

$$L = \int_\alpha^\beta \sqrt{-g(\dot{\gamma}, \dot{\gamma})} d\lambda. \qquad (A.136)$$

Except for the choice $g(\dot{\gamma}, \dot{\gamma}) = -1$, the calculation is essentially identical and the same result, $\nabla_{\dot{\gamma}} \dot{\gamma} = 0$, emerges. However, if γ is null (i.e. $g(\dot{\gamma}, \dot{\gamma}) = 0$) then a different approach must be taken.

7.7. Since $\nabla_X(fY) = (Xf)Y + f\nabla_X Y$ and $[X, fY] = f[X, Y] + (Xf)Y$, we see $\nabla_X Y - [X, Y]$ behaves as a tensor in its second argument. The identity $\nabla_{fY} X = f\nabla_Y X$ is a fundamental property of a connection, and so, using Eq. (7.27), we have the first required result $T(X, fY) = fT(X, Y)$. Combining the first result with the identity $T(X, Y) = -T(Y, X)$ gives the second required result $T(fX, Y) = fT(X, Y)$.

7.8. Using Eq. (7.27), the components of the torsion tensor are given by

$$
\begin{aligned}
T^a{}_{bc} &= e^a\left(T(X_b, X_c)\right) \\
&= e^a(\nabla_{X_b} X_c) - e^a(\nabla_{X_c} X_b) - e^a([X_b, X_c]) \\
&= \Gamma^a{}_{bc} - \Gamma^a{}_{cb} - f_{bc}{}^a.
\end{aligned} \qquad (A.137)
$$

If the connection is torsion-free (hence $T^a{}_{bc} = 0$) then, by inspection, $\Gamma^a{}_{bc} = \Gamma^a{}_{cb}$ if the structure coefficients satisfy $f_{bc}{}^a = 0$. The structure coefficients vanish if the basis is a coordinate frame, i.e. $\{X_a = \partial/\partial x^a\}$ in some coordinate system $\{x^a\}$.

7.9. The covariant derivative commutes with tensor contractions, and so

$$\nabla_{\dot\gamma}(g(V, W)) = (\nabla_{\dot\gamma}g)(V, W) \\ + g(\nabla_{\dot\gamma}V, W) + g(V, \nabla_{\dot\gamma}W). \tag{A.138}$$

So, although $\nabla_{\dot\gamma}V = \nabla_{\dot\gamma}W = 0$, we see that $\nabla_{\dot\gamma}g \neq 0$ can yield $\nabla_{\dot\gamma}(g(V, W)) \neq 0$. A similar argument applies to $\nabla_{\dot\gamma}(g(V, V))$ and $\nabla_{\dot\gamma}(g(W, W))$, and thus it also applies to θ.

Since $\nabla_{\dot\gamma}V = \nabla_{\dot\gamma}W = 0$ we find

$$\nabla_{\dot\gamma}(\cos\theta) = \frac{(\nabla_{\dot\gamma}g)(V, W)}{\sqrt{g(V, V)g(W, W)}} - \frac{1}{2}\frac{(\nabla_{\dot\gamma}g)(V, V)\, g(V, W)}{\sqrt{g(V, V)^3 g(W, W)}} \\ - \frac{1}{2}\frac{(\nabla_{\dot\gamma}g)(W, W)\, g(V, W)}{\sqrt{g(V, V)g(W, W)^3}}, \tag{A.139}$$

using the Leibniz rule, the chain rule and (A.138). The non-metricity is such that $\nabla_X g = \alpha(X)g$; thus $\nabla_{\dot\gamma}(\cos\theta) = 0$, i.e. θ is constant along γ.

7.10. The structure coefficients of any coordinate frame vanish. Furthermore, $Q = 0$ and $T = 0$ so Eq. (7.32) reduces to

$$2g(\nabla_{\partial_a}\partial_b, \partial_c) = \partial_a[g(\partial_b, \partial_c)] + \partial_b[g(\partial_c, \partial_a)] \\ - \partial_c[g(\partial_a, \partial_b)]. \tag{A.140}$$

In other words,

$$2g_{cd}\,\Gamma^d{}_{ab} = \partial_a g_{bc} + \partial_b g_{ca} - \partial_c g_{ab}. \tag{A.141}$$

Eq. (7.23) then emerges by making use of $g^{ac}g_{cb} = \delta^a_b$ and $g_{ab} = g_{ba}$, with the appropriate relabelling of indices.

7.11. Since

$$\nabla_X\nabla_Y(fZ) = \nabla_X((Yf)Z + f\nabla_Y Z) \\ = (XYf)Z + (Yf)\nabla_X Z \\ + (Xf)\nabla_Y Z + f\nabla_X\nabla_Y Z \tag{A.142}$$

it follows

$$\nabla_X \nabla_Y (fZ) - \nabla_Y \nabla_X (fZ) = f(\nabla_X \nabla_Y Z - \nabla_Y \nabla_X Z)$$
$$+ ([X,Y]f)Z. \qquad (A.143)$$

Furthermore, $\nabla_{[X,Y]}(fZ) = ([X,Y]f)Z + f\nabla_{[X,Y]}Z$ and hence

$$\nabla_X \nabla_Y (fZ) - \nabla_Y \nabla_X (fZ) - \nabla_{[X,Y]}(fZ)$$
$$= f(\nabla_X \nabla_Y Z - \nabla_Y \nabla_X Z - \nabla_{[X,Y]}Z) \qquad (A.144)$$

i.e. $R(X,Y)(fZ) = fR(X,Y)(Z)$. The final required result follows in a similar manner.

7.12. The corresponding mixed tensor is $e^a \otimes X_a$ because $(e^a \otimes X_a)(V,-) = e^a(V)X_a$, where the dash indicates an unevaluated argument. However, $e^a(V)X_a = V$ so $e^a \otimes X_a$ is the identity $(1,1)$-tensor. Notice that we could have introduced $X_a \otimes e^a$ instead of $e^a \otimes X_a$, but left the first argument unevaluated when acting on V.

7.13. Begin by noting $\underset{\sim}{e^b} = \widetilde{X^b}$ where $X^b = g^{bc}X_c$. The Hodge map satisfies $\star(\widetilde{X_a} \wedge \widetilde{X^b}) = i_{X^b} \star \widetilde{X_a}$ and, bearing in mind $R^a{}_b$ is a 2-form,

$$R^a{}_b \wedge i_{X_b} \star \widetilde{X_a} = i_{X^b}(R^a{}_b \wedge \star\widetilde{X_a}) - i_{X^b}R^a{}_b \wedge \star\widetilde{X_a} \qquad (A.145)$$

using the graded Leibniz rule for the interior product. However, $R^a{}_b \wedge \star\widetilde{X_a}$ is a $(n+1)$-form, where n is the dimension of the manifold, and it is therefore zero. A similar argument can be used to express the remaining term in (A.145) as $-i_{X_a}i_{X^b}R^a{}_b \star 1$ (or one can simply use $\alpha \wedge \star\widetilde{X} = \alpha(X) \star 1$, where α is any 1-form and X is any vector). Hence, we arrive at

$$R^a{}_b \wedge \star(\widetilde{X_a} \wedge \underset{\sim}{e^b}) = i_{X^b}i_{X_a}R^a{}_b \star 1 \qquad (A.146)$$

where the antisymmetry of the interior product (to be specific, $-i_{X_a}i_{X^b} = i_{X^b}i_{X_a}$) has been used. However, $i_Y i_X \beta = 2\beta(X,Y)$ where β is any 2-form and X, Y are any vectors. Hence, Eq. (7.40) can be expressed as

$$R(X_a, X^b) = i_{X^b}i_{X_a}R^c{}_d e^d \otimes X_c \qquad (A.147)$$

which is equivalent to

$$e^c(R(X_a, X^b)X_d) = i_{X^b}i_{X_a}R^c{}_d. \qquad (A.148)$$

On the other hand, Eq. (7.34) and Eq. (7.35) together yield

$$\mathcal{R} = e^a(R(X_a, \tilde{e}^b)X_b) \qquad (A.149)$$

and so $\mathcal{R} = i_{X^b}i_{X_a}R^a{}_b$. The required result follows immediately from (A.146).

7.14. Acting with the exterior derivative on $g^{ac}g_{cb} = \delta^a_b$ gives $dg^{ac}g_{cb} + g^{ac}dg_{cb} = 0$ because the Kronecker delta is constant (it is 1 or 0 depending on whether or not the indices are equal). Thus, $dg^{ab} = -g^{ac}g^{bd}dg_{cd}$ and it follows

$$dg^{ab} \wedge dg_{ab} = -g^{ac}g^{bd}dg_{cd} \wedge dg_{ab}. \qquad (A.150)$$

However, introducing the relabelling $a \leftrightarrow c$ and $b \leftrightarrow d$ and making use of the antisymmetry of the wedge product yields $g^{ac}g^{bd}dg_{cd} \wedge dg_{ab} = -g^{ca}g^{db}dg_{cd} \wedge dg_{ab}$. Thus, we conclude $dg^{ab} \wedge dg_{ab} = -dg^{ab} \wedge dg_{ab}$, and the desired result is found.

7.15. Since $\mathfrak{F}' = \mu i_{\dot\gamma} \star F$, we see that $\mathfrak{F}'$ can be obtained by making the replacements $q \mapsto \mu$ and $F \mapsto \star F$ in the Lorentz force 1-form Eq. (7.61). Comparison of Eq. (5.15) and Eq. (5.31) shows that the latter replacement corresponds to $\mathbf{E} \mapsto -\mathbf{B}$ and $\mathbf{B} \mapsto \mathbf{E}$. Thus, $\mathfrak{F}'$ is the force on a hypothetical magnetically-charged point particle, i.e. a magnetic monopole. The particle has magnetic charge $-\mu$.

7.16. The Fermi-Walker derivative $\nabla^F_{\dot\gamma}$ commutes with contractions so

$$\nabla^F_{\dot\gamma}(g(X,Y)) = (\nabla^F_{\dot\gamma}g)(X,Y) \\ + g(\nabla^F_{\dot\gamma}X, Y) + g(X, \nabla^F_{\dot\gamma}Y). \qquad (A.151)$$

However, $\nabla^F_{\dot\gamma}f = \dot\gamma(f) = \nabla_{\dot\gamma}f$, where f is any 0-form, and so

$$\nabla^F_{\dot\gamma}(g(X,Y)) = g(\nabla_{\dot\gamma}X, Y) + g(X, \nabla_{\dot\gamma}Y) \qquad (A.152)$$

since $\nabla g = 0$. On the other hand, using Eq. (7.70), the terms involving the acceleration in $\nabla^F_{\dot\gamma}X$ and $\nabla^F_{\dot\gamma}Y$ cancel in $g(\nabla^F_{\dot\gamma}X, Y) + g(X, \nabla^F_{\dot\gamma}Y)$ giving

$$g(\nabla^F_{\dot\gamma}X, Y) + g(X, \nabla^F_{\dot\gamma}Y) = g(\nabla_{\dot\gamma}X, Y) + g(X, \nabla_{\dot\gamma}Y). \qquad (A.153)$$

Hence, combining (A.151), (A.152) and (A.153) yields the required result.

7.17. The vector S is spacelike, so the square of its magnitude is $||S||^2 = g(S,S)$. Clearly, $||S||$ is constant if $g(S,S)$ is constant. The Fermi-Walker derivative of $g(S,S)$ is

$$\nabla^F_{\dot\gamma}(g(S,S)) = 2g(S,\nabla^F_{\dot\gamma}S) = 2i_S\nabla^F_{\dot\gamma}\widetilde{S} \qquad (A.154)$$

because $\nabla^F_{\dot\gamma}$ is compatible with the metric, and the metric is symmetric. However, $i_S(\widetilde{\dot\gamma}\wedge i_S F) = 0$ because S is orthogonal to $\dot\gamma$ (so $i_S\widetilde{\dot\gamma} = 0$) and $i_S i_S = 0$ using the rules of exterior calculus. The antisymmetry of the interior product also ensures $i_S i_{\dot\gamma} = -i_{\dot\gamma} i_S$, and thus the action of i_S on the right-hand side of Eq. (7.76) yields zero. Hence, substituting Eq. (7.76) in (A.154) leads us to conclude $g(S,S)$ is constant along γ.

7.18. Since S is orthogonal to $\dot\gamma$, and the Fermi-Walker derivative is compatible with the metric, the metric dual of the left-hand side of Eq. (7.76) reduces to

$$\nabla^F_{\dot\gamma}S = \nabla_{\dot\gamma}S - \frac{q}{m}i_S i_{\dot\gamma}F\,\dot\gamma \qquad (A.155)$$

using Eq. (7.70) and the given expression for the acceleration. Furthermore, $i_{\dot\gamma}(\widetilde{\dot\gamma}\wedge i_S F) = i_{\dot\gamma}\widetilde{\dot\gamma}i_S F - i_{\dot\gamma}i_S F\widetilde{\dot\gamma} = -i_S F + i_S i_{\dot\gamma}F\widetilde{\dot\gamma}$ using $g(\dot\gamma,\dot\gamma) = -1$ and the antisymmetry of the interior product. Substituting this result in the right-hand side of Eq. (7.76), using $\widetilde{\nabla^F_{\dot\gamma}\widetilde{S}} = \nabla^F_{\dot\gamma}S$ and collecting the coefficients of $\dot\gamma$ leads to the required result.

A.8 GENERALISED FUNCTIONS FROM A GEOMETRIC PERSPECTIVE

8.1. Begin by noting $\alpha\wedge\omega$ has degree $p+q = n+1$, which is greater than the dimension of the manifold. Thus $\alpha\wedge\omega = 0$. However, $i_Y(\alpha\wedge\omega) = i_Y\alpha\wedge\omega + (-1)^q\alpha\wedge i_Y\omega$ using the graded Leibniz rule and so $i_Y\alpha\wedge\omega = -(-1)^q\alpha\wedge i_Y\omega$. However, the distribution α_D satisfies $\alpha_D[i_Y\omega] = \int_M \alpha\wedge i_Y\omega$ and therefore $\alpha_D[i_Y\omega] = -(-1)^q\int_M i_Y\alpha\wedge\omega = -(-1)^q(i_Y\alpha)_D[\omega]$. Identifying $i_Y\alpha_D$ with $(i_Y\alpha)_D$ yields the required result.

8.2. The distribution $\mathcal{I}_D$ has degree 0, and so $d\mathcal{I}_D[f] = -\mathcal{I}_D[df]$. Substituting the result $\mathcal{I}_D[df] = f(x_2) - f(x_1)$ found in the main text immediately gives the required result.

8.3. Begin by noting $\int_{\partial \mathcal{U}} dr \wedge \chi = 0$ because χ vanishes on the end caps of $\partial \mathcal{U}$ and r is constant elsewhere on $\partial \mathcal{U}$; thus $\partial \mathcal{U}_D[dr \wedge \chi] = 0$. However, by definition, $(\partial \mathcal{U}_D \wedge dr)[\chi] = \partial \mathcal{U}_D[dr \wedge \chi]$ and we see $(\partial \mathcal{U}_D \wedge dr)[\chi] = 0$. Hence, $\partial \mathcal{U}_D \wedge dr = 0$ holds on the class of test 1-forms whose support does not intersect with the end caps of $\partial \mathcal{U}$.

8.4. The distribution $\partial \mathcal{U}_D$ has degree 1, and so the graded Leibniz rule gives $i_{\partial_t}(\partial \mathcal{U}_D \wedge dr) = i_{\partial_t}\partial \mathcal{U}_D\, dr - i_{\partial_t} dr\, \partial \mathcal{U}_D$. However, $i_{\partial_t} dr = \partial_t r = 0$ and thus the first required result holds because $\partial \mathcal{U}_D \wedge dr = 0$. The working towards the second required result is identical, except $\partial/\partial t$ is replaced by $\partial/\partial \theta$.

8.5. Since $\mathcal{U}$ is a 3-dimensional region, the boundary of $\mathcal{U}$ is 2-dimensional. Thus, the distribution $\partial \mathcal{U}_D = (\partial \mathcal{U})_D$ acts on test 2-forms and, since the spacetime is 4-dimensional, it follows that the degree of $\partial \mathcal{U}_D$ is $4 - 2 = 2$. Any 2-distribution can be expressed as $\Theta_D = \frac{1}{2}\Theta_{Dab}\, dx^a \wedge dx^b$ where the coordinates are chosen such that $x^2 = \xi$ and $x^3 = \zeta$. Inspection of each term in Θ_D shows that it has the general structure given in the hint in the question. Now note $dx^0 \wedge d\xi$, $dx^1 \wedge d\xi$, $dx^0 \wedge d\zeta$ and $dx^1 \wedge d\zeta$ are all non-zero (otherwise $\{x^0, x^1, x^2, x^3\}$ would not be a valid coordinate system). However, $\partial \mathcal{U}_D \wedge d\zeta = 0$ and so the components of terms that do not contain $d\zeta$ must be zero. Likewise, $\partial \mathcal{U}_D \wedge d\xi = 0$ ensures terms that do not contain $d\xi$ must also be zero. It follows that $\partial \mathcal{U}_D$ must have the structure stated in the question.

8.6. The 2-dimensional region $\Sigma^{(1)}$ swept out in spacetime by the inner edge is part of the boundary of the 3-dimensional spacetime region swept out by the hole in the sample. Since $dj^{(1)} \simeq \check{\sigma} F$ on the inner edge, following similar working to that used in the main text leads to $\Delta Q^{(1)} = -\check{\sigma}\Delta \Phi^{(1)}$ where $\Delta Q^{(1)}$ is the change in the charge in the inner edge and $\Delta \Phi^{(1)}$ is the change in the magnetic flux through the hole. Likewise, the 2-dimensional region $\Sigma^{(2)}$ swept out in spacetime by the outer edge is part of the boundary of the 3-dimensional spacetime region swept out by the union of the sample and the hole. Since $dj^{(2)} \simeq -\check{\sigma} F$ on the outer edge, we find $\Delta Q^{(2)} = \check{\sigma}\Delta \Phi^{(2)}$ must hold for the change $\Delta Q^{(2)}$ in the charge in the outer edge and the change $\Delta \Phi^{(2)}$ in the magnetic flux through the union of the sample and the hole. If the magnetic flux through the bulk of the sample is zero then $\Delta \Phi^{(1)} = \Delta \Phi^{(2)}$ and we find $\Delta Q^{(2)} = -\Delta Q^{(1)}$ as asserted in part (a). For part (b), note that the relationship between the charges in the edges must hold throughout the deformation. The geometry of the sample does not play a role in the calculation in part (a). We conclude that the sign of the charge in the inner edge before the deformation must be opposite to the sign of the charge in the inner edge after the deformation. Likewise, the sign of the charge in the outer edge before the deformation must be opposite to the sign of the charge in the outer edge after the deformation. It is clear from this thought experiment that the upper and lower faces of the sample are distinguishable, which is a reflection of the fact that $\mathcal{U}$ is an orientable 3-dimensional submanifold.

8.7. Begin by noting that Eq. (8.69) yields

$$\star(\gamma_D \alpha)[\omega] = (-1)^{p(n-p)}(\gamma_D \alpha)[\star\omega] \tag{A.156}$$

where ω is any test p-form. Furthermore,

$$\begin{aligned}(\gamma_D \alpha)[\star\omega] &= \gamma_D[\alpha \wedge \star\omega] \\ &= \gamma_D[\omega \wedge \star\alpha]\end{aligned}$$

using Eq. (8.12) in the first step and the symmetry of the metric on p-forms in the second step. However, $\omega \wedge \star\alpha = (-1)^{p(n-p)}\star\alpha \wedge \omega$ because ω is a p-form and $\star\alpha$ is a $(n-p)$-form. Thus $\star(\gamma_D \alpha)[\omega] = \gamma_D[\star\alpha \wedge \omega] = (\gamma_D \star \alpha)[\omega]$ as required.

8.8. The result

$$\begin{aligned}\star F = {}&-B_x\, dt \wedge dx - B_y\, dt \wedge dy - B_z\, dt \wedge dz \\ &- E_x\, dy \wedge dz - E_y\, dz \wedge dx - E_z\, dx \wedge dy\end{aligned} \tag{A.157}$$

proven in Exercise 5.4 leads to

$$\begin{aligned}dz \wedge \star F = {}&-B_x\, dz \wedge dt \wedge dx - B_y\, dz \wedge dt \wedge dy \\ &- E_z\, dz \wedge dx \wedge dy.\end{aligned} \tag{A.158}$$

and so $dz \wedge \star(F^{\mathrm{II}} - F^{\mathrm{I}}) = dz \wedge j^{\Sigma}$ yields

$$\check{\rho} = E_z^{\mathrm{II}} - E_z^{\mathrm{I}}, \quad \check{j}_x = -B_y^{\mathrm{II}} + B_y^{\mathrm{I}}, \quad \check{j}_y = B_x^{\mathrm{II}} - B_x^{\mathrm{I}} \tag{A.159}$$

i.e. $\check{\rho} = \mathbf{e}_z \cdot (\mathbf{E}^{\mathrm{II}} - \mathbf{E}^{\mathrm{I}})$ and $\check{\mathbf{j}} = \mathbf{e}_z \times (\mathbf{B}^{\mathrm{II}} - \mathbf{B}^{\mathrm{I}})$ as required in the first part of the exercise.

The second part of the exercise yields the same conclusion because $dz \simeq 0$. We see

$$\star F \simeq -B_x \, dt \wedge dx - B_y \, dt \wedge dy - E_z \, dx \wedge dy \qquad (A.160)$$

using (A.157), and inspection of $\star(F^{II} - F^{I}) \simeq j^{\Sigma}$ yields the required result.

8.9. The inverse metric tensor G_1 is

$$G_1 = -\frac{\partial}{\partial t} \otimes \frac{\partial}{\partial t} + \frac{\partial}{\partial x} \otimes \frac{\partial}{\partial x} + \frac{\partial}{\partial y} \otimes \frac{\partial}{\partial y} + \frac{\partial}{\partial z} \otimes \frac{\partial}{\partial z}, \qquad (A.161)$$

and so $G_1(\mathbf{e}, \mathbf{e}) = -1$, i.e. $\mathbf{e}$ is timelike and unit normalised. Furthermore, $G_1(d\lambda, \mathbf{e}) = 0$ follows from $d\lambda = dx - \dot{X} \, dt$, i.e. $\mathbf{e}$ is orthogonal to $d\lambda$.

Since $\star(dx^+ \wedge dy) = dx^+ \wedge dz$ and $\star(dx^- \wedge dy) = -dx^- \wedge dz$ we have

$$\star F = -f_{\text{in}}(x^-) \, dx^- \wedge dz + f_{\text{out}}(x^+) \, dx^+ \wedge dz \qquad (A.162)$$
$$\simeq -2 f_{\text{in}}(t - X(t))(1 - \dot{X}) \, dt \wedge dz \qquad (A.163)$$

using Eq. (8.77) and the result of $F \simeq 0$ found in the main text. However,

$$\mathbf{e} \simeq \frac{\dot{X}^2 - 1}{\sqrt{1 - \dot{X}^2}} \, dt$$
$$= -\sqrt{1 - \dot{X}^2} \, dt \qquad (A.164)$$

and so, using $j^{\Sigma} \simeq -\star F$, we obtain

$$j^{\Sigma} \simeq -2 f_{\text{in}}(t - X(t)) \sqrt{\frac{1 - \dot{X}}{1 + \dot{X}}} \, \mathbf{e} \wedge dz. \qquad (A.165)$$

The previous result is equivalent to the equality

$$d\lambda \wedge j^{\Sigma} = -2 f_{\text{in}}(t - X(t)) \sqrt{\frac{1 - \dot{X}}{1 + \dot{X}}} \, d\lambda \wedge \mathbf{e} \wedge dz, \qquad (A.166)$$

and we see that j_{Σ} can only differ from the result stated in the question by a 2-form with structure $d\lambda \wedge \alpha$. The 1-form α can be chosen such that $\alpha(Y) = 0$ without loss of generality, and we find

$i_Y \mathsf{j}^\Sigma = g(Y,Y)\,\alpha$. However, α must be zero if $i_Y \mathsf{j}^\Sigma = 0$ because $g(Y,Y) \neq 0$. We conclude that the equality after pullback satisfied by j^Σ is in fact just an equality (at all points on Σ).

The vector field $\widetilde{\mathbf{e}}$ is unit normalised and timelike. It is tangent to the worldlines of observers who are moving at the same velocity as the mirror. Since $\mathbf{e}$ reduces to $-dt$ when the mirror is at rest, we see that the $\mathbf{e} \wedge dz$ component of the surface current 2-form is the negative of the y-component of the surface current in the frame of those observers.

Suggestions for further reading

The following brief list is only indicative. It is by no means exhaustive.

Arnold, V. I. (1978). *Mathematical Methods of Classical Mechanics*. New York: Springer-Verlag.

Benn, I. M. and Tucker, R. W. (1989). *An Introduction to Spinors and Geometry with Applications in Physics*. Bristol: Adam Hilger.

Crampin, M. and Pirani, F. A. E. (1987). *Applicable Differential Geometry*. Cambridge: Cambridge University Press.

De Rham, G. (1984). *Differentiable Manifolds*. Heidelberg: Springer Berlin.

Fecko, M. (2006). *Differential Geometry and Lie Groups for Physicists*. Cambridge: Cambridge University Press.

Flanders, H. (1989). *Differential Forms with Applications to the Physical Sciences*. New York: Dover Publications.

Frankel, T. (2011). *The Geometry of Physics: an Introduction*. Cambridge: Cambridge University Press.

Holm, D. D. (2011). *Geometric Mechanics*. London: Imperial College Press.

Isham, C. J. (1999). *Modern Differential Geometry for Physicists*. Singapore: World Scientific.

Schutz, B. F. (1980). *Geometrical Methods of Mathematical Physics*. Cambridge: Cambridge University Press.

Suggestions for further reading

Index

acceleration, 115
action, 80, 88, 94
almost symplectic structure, 98
Ampère's law, 21, 23, 25
angle, 39, 50, 109
arc-length, 37
autoparallel, 104, 117

basis
    1-form, 17
    p-form, 17
    Poisson bracket, 91
    vector, 10, 18
boundary, 22, 28, 70, 126, 135

Cartan
    identity, 53, 90, 119
    structure equations, 113
causal structure, 51
charge conservation, 68
coframe, 20
    orthonormal, 34, 35
compact support, 124
conformal structure, 50
connection, 101
    1-forms, 101
    coefficients, 101, 108
conserved quantity, 81, 92
constitutive equations, 74
contraction, 12, 15, 19, 20
cotangent
    bundle, 84
    space, 34, 84
covariant differentiation, 101

covector, 3, 15
current conservation, 133
curvature scalar, 112
curve
    length, 43, 105
    lightlike, see null
    null, 38
    spacelike, 38
    timelike, 38

differential form, 1, 2, 15, 17
    0-form, 4, 17
    1-form, 2, 11, 17
    p-form, 6, 17
    closed, 18
    degree, 6, 17
    exact, 18
directional derivative, 10
distribution, 125
    degree, 127
divergence theorem, 27

Einstein summation, 33
electric current 3-form, 67, 73
electromagnetic
    2-form, 59, 73, 118
    invariants, 62
equality after pullback, 138
Euler-Lagrange equations, 80
exterior calculus, 4
exterior derivative, 4, 7, 18, 23, 126
    graded Leibniz rule, 7, 128
exterior product, see wedge product

Fermi-Walker derivative, 120, 122
first law of thermodynamics, 3, 12
 constant pressure, 11
 constant volume, 11
 heat capacity, 10, 12, 16
 internal energy, 12, 14
 temperature, 10
frame, 20
 orthonormal, 33

Gauss's law, 26
Gauss-Ampère law, 61, 142
 macroscopic, 74
Gauss-Faraday law, 60, 132, 142
generalised function, 124
generalised Stokes theorem, 27,
 126
geodesic, 107
gravitational force, 8

Hamilton's equations, 88
Hamiltonian, 88, 95
Hamiltonian vector field, 89
Hodge map, 47, 51, 57, 61, 144

inner product, 47, 48
interior product, 15, 20, 128
 graded Leibniz rule, 20

Jacobi identity, 91, 93, 99
junction condition, 143

Killing vector, 54

Lagrangian, 80, 87
Laplace equation, 51
length, 37
Levi-Civita connection, 105, 106,
 115
Lie derivative, 53, 89, 108
lift, 78

Lorentz force, 84, 118

magnetisation, 72
magnetostatics, 21
magnitude of a vector, 30
manifold, 76
Maxwell 2-distribution, 142
Maxwell 2-form, see
 electromagnetic 2-form
Maxwell relations, 5
metric
 dual, 46, 57
 Euclidean, 36
 induced, 42–45
 inverse, 47
 Lorentzian, 36
 signature, 36
 tensor, 9, 30, 57, 108, 110

non-degenerate
 2-form, 86, 98
 metric, 32, 33, 47
non-metricity, 108, 110, 114
null vector, 51

orientation, 26, 70, 135

parallel transport, 103, 108, 121
Poincaré 1-form, 85
Poisson bracket, 91
polarisation, 72
potential flow, 51
proper time, 38

rapidity, 39, 50
Ricci
 2-form, 113, 115
 tensor, 112
Riemann curvature tensor, 111,
 114

scalar force, 115

scalar lift, 79
scalar product, 34
section, 78
singular Lagrangian, 87, 94
spacelike vector, 51
spin vector, 120, 122
Stokes's theorem, 22, 24
structure coefficients, 108
submanifold, 76, 93, 95
support, 125
symplectic 2-form, 85, 89, 90

tangent
    bundle, 77
    space, 34, 77
    vector, 37
tensor, 30
    product, 31
test
    form, 125
    function, 124
timelike vector, 51
torsion, 107, 110, 113

vector field, 8–10, 14, 15, 18, 77
volume, 41
    n-form, 41

wedge product, 4, 6, 17, 127
worldline, 38

Printed in the United States
by Baker & Taylor Publisher Services